War, Terrible War

A HISTORY OF US

BOOK ONE The First Americans
BOOK TWO Making Thirteen Colonies
BOOK THREE From Colonies to Country
BOOK FOUR The New Nation
BOOK FIVE Liberty for All?
BOOK SIX War, Terrible War
BOOK SEVEN Reconstruction and Reform
BOOK EIGHT An Age of Extremes
BOOK NINE War, Peace, and All That Jazz
BOOK TEN All the People

Oxford University Press

OXFORD
A HISTORY OF
US
BOOK SIX

War, Terrible War

Joy Hakim

Oxford University Press
New York

Oxford University Press
Oxford   New York
Athens   Auckland   Bangkok   Bombay
Calcutta   Cape Town   Dar es Salaam   Delhi
Florence   Hong Kong   Istanbul   Karachi
Kuala Lumpur   Madras   Madrid   Melbourne
Mexico City   Nairobi   Paris   Singapore
Taipei   Tokyo   Toronto

and associated companies in
Berlin   Ibadan

Designer: Mervyn E. Clay
Maps copyright © 1994 by Wendy Frost and Elspeth Leacock
Produced by American Historical Publications

Published by Oxford University Press, Inc.
200 Madison Avenue, New York, New York 10016
Oxford is a registered trademark of Oxford University Press

Library of Congress Cataloging-in-Publication Data
Hakim, Joy.
War, terrible war / Joy Hakim.
p.   cm.—(A history of US: bk. 6)
Includes bibliographical references and index.
ISBN 0-19-507755-5 (lib. ed.)—ISBN 0-19-507765-2 (series, lib. ed.)
ISBN 0-19-507756-3 (paperback ed.)—ISBN 0-19-507766-0 (series, paperback ed.)
ISBN 0-19-509511-1 (trade hardcover ed.)—ISBN 0-19-509484-0 (series, trade hardcover ed.)
1. United States—History—Civil War, 1861–1865—Juvenile literature.
[1. United States—History—Civil War, 1861–1865.] I. Title. II. Series: Hakim, Joy. History of US; 6.
E178.3.H22 1994 vol. 6
[E468]
973.7—dc20   93-26252
CIP
AC

3 5 7 9 8 6 4
Printed in the United States of America
on acid-free paper

I cannot give you an idea of the terrors of this battle. I believe it was as hard a contested battle as was ever fought on the American continent, or perhaps anywhere else....For ten long hours...the firing did not cease for a moment. Try to picture...one hundred thousand men, all loading and firing as fast as they could....The sight of the dead, the cries of the wounded, the thundering noise of the battle, can never be put on paper....The dead, the dying and the wounded...all mixed up together; friend and foe embraced in death; some crying for water; some praying their last prayers; some trying to whisper to a friend their last farewell message to their loved ones at home. It is heartrending. I cannot go any further. Mine eyes are damp with tears....Although the fight is over the field is yet quite red with blood....The victory was complete. We are now occupying the same ground that we did before the battle.

—Private J. W. Reid, 4th regiment, South Carolina Volunteers

The President came into the office laughing with a volume of comedy in his hand, seemingly utterly unconscious that he, with his shortshirt hanging about his long legs and setting out behind like the tail-feathers of an enormous ostrich, was infinitely funnier than anything in the book he was laughing at. What a man it is.

—Lincoln's secretary John Hay, in his diary for May 14, 1864

Whatever may be the result of the contest I foresee that the country will have to pass through a terrible ordeal...for our national sins.

—R. E. Lee, general, C.S.Army

The art of war is simple enough. Find out where your enemy is. Get at him as soon as you can. Strike at him as hard as you can and as often as you can, and keep moving on.

—U. S. Grant, general, U.S. Army

This is our native country; we have as strong attachment naturally to our native hills, valleys, plains, luxuriant forest, flowing streams, mighty rivers, and lofty mounts, as any other people....We love this land, and have contributed our share to its prosperity and wealth.

—John Swett Rock, the first black lawyer admitted to practice before the U.S. Supreme Court

Contents

PREFACE I: Dinner at Brown's Hotel   9

PREFACE II: A Divided Nation   11

1 Americans Fighting Americans   14

2 The War Begins   17

3 Harriet and *Uncle Tom*   23

4 Harriet, Also Known as Moses   27

FEATURE: BREAKING THE LAW: A DISCUSSION OF ETHICS   32

5 Abraham Lincoln   34

6 New Salem   38

7 Mr. President Lincoln   41

8 President Jefferson Davis   45

9 Slavery   48

FEATURE: SLAVERY: THEN AND NOW   50

10 John Brown's Body   54

11 Lincoln's Problems   59

12 The Union Generals   64

13 The Confederate Generals   68

14 President Davis's Problems   73

15 Choosing Sides   76

16 The Soldiers   80

17 Willie and Tad   86

18 General McClellan's Campaign   89

19 War at Sea   94

FEATURE: RULER OF THE PRESIDENT'S NAVY   97

20 Emancipation Means Freedom   98

21 Determined Soldiers — **103**

22 Marching Soldiers — **107**

23 Awesome Fighting — **111**

24 Lee the Fox — **117**

25 Speeches at Gettysburg — **119**

26 More Battles—Will It Ever End? — **124**

    FEATURE: THE TURN OF THE TIDE — **129**

27 The Second Inaugural — **130**

28 Closing In on the End — **133**

29 Mr. McLean's Parlor — **138**

30 A Play at Ford's Theatre — **143**

31 After Words — **147**

    FEATURE: SONGS OF THE CIVIL WAR — **150**

    CHRONOLOGY OF EVENTS — **153**

    MORE BOOKS TO READ — **154**

    INDEX — **155**

    PICTURE CREDITS — **159**

    A NOTE FROM THE AUTHOR — **160**

General Robert E. Lee

*I*n this book you may notice that some words are not spelled exactly the way you have learned to spell them. Spelling in America, like everything else, has changed somewhat since the 19th century. Words such as **theater** and **traveler** were spelled **theatre** and **traveller**— usually because that was the way they were spelled in Britain. It took this country longer to break away from British spelling than to break away from the British government.

PREFACE I
Dinner at Brown's Hotel

A preface is an explanation that comes before a book gets started and tells what it is about. This book is about war—CIVIL WAR—which is war inside a nation, and that takes some powerful explaining. Our Civil War takes so much explaining that this book needed **two** *prefaces.*

THE TIME: *April 13, 1830*
THE PLACE: *Brown's Indian Queen Hotel, Washington, D.C.*
THE OCCASION: *A dinner to celebrate Thomas Jefferson's birthday* [Jefferson had been dead four years, but politicians like dinners and celebrations.]
THE MAIN CHARACTERS: *Vice President John C. Calhoun and President Andrew Jackson*

The nation was in trouble. Only a few people could see that, but President Andrew Jackson was one of those who was alarmed. The country— which he and everyone else called "the Union," or the "Federal Union"—was being divided in half, right along its belt line. It was being divided into North and South, and those terms had to do with much more than geography. People in the North and South were beginning to dislike each other, and say so with angry words.

There was that old problem of what Southerners called their "peculiar institution," and others called "slavery." Slavery was making people in the North and South think and act differently from each other.

There were other problems. The old southern states, like South

"I would rather die in the last Ditch," said Jackson, than "see the Union disunited."

You'll soon be reading about another Jackson, called Stonewall. Don't confuse the two Jacksons; they are different men.

John Calhoun (above) and Andrew Jackson grew further and further apart until Calhoun resigned as Jackson's vice president.

9

Which were the "old" southern states? Which southern states were "new"?

Carolina, were having economic troubles. Their land was worn out. South Carolina blamed the North, and some of Congress's laws, for its problems.

John Calhoun—South Carolina's handsome but unsmiling political leader—said each state had the right to decide if Congress's laws were constitutional or not. That meant each state could decide which laws to obey, and which not. (You can see that the nation wouldn't last long if each state made its own rules.)

South Carolina was used to having its way. It was the state that, 43 years earlier, had insisted that slavery be allowed in the new nation's constitution. Now it was insisting on what were called *states' rights*. And that was what most of the toasts were about at that dinner at Brown's Indian Queen Hotel.

There were 24 toasts that evening before they got to Andrew Jackson. People could see that he was fuming over the ideas being toasted. The president rose to his feet—and so did everyone else. Martin Van Buren (who was soon to be president) climbed on a chair so he could see. Then the man they called "Old Hickory" raised his glass high above his silver hair, glared at John Calhoun, and said:

"Our Union—it must be preserved."

They say that Calhoun's hand shook and some wine trickled down his arm, but he was undaunted. He gave his toast.

"The Union—next to our liberty, most dear."

What happened next? People in the North heard Jackson's message, but in the South they listened to John Calhoun. Calhoun told them that slavery was a "positive good" and that a belief in states' rights was a belief in liberty.

That conflict of ideas would create the worst war in all our history. Go on to the next page for more background. You'll need it. There's much adventure, turmoil, and bloodshed ahead of you in this book.

Calhoun's idea of states' rights went back to the time of the American Revolution and to Patrick Henry. Henry said he "smelled a rat" in the new Constitution. The rat was a strong central government. Calhoun and Henry felt that the states should be more powerful than the federal government.

Toasts are little speeches that people sometimes give before drinking. They have nothing to do with burned bread. Well, that is not quite true. In Shakespeare's day a small piece of toasted bread was often put in a wineglass. It was a way of collecting the impurities in the wine at the bottom of the glass.

Supporters of slavery said that slaves needed the system as much as the owners did. This cartoonist disagreed.

PREFACE II
A Divided Nation

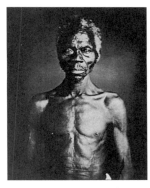

This slave was photographed for a Harvard scientist who wanted to prove that blacks were physically inferior to whites. What he did prove was that scientists can be as prejudiced as anyone else.

We hold these truths to be self-evident, that all men are created equal, that they are endowed by their creator with certain unalienable rights, that among these are life, liberty and the pursuit of happiness.

The time was 1860, and Americans had a problem. It wasn't a new problem; they'd been living with it since the nation began. There were those words in the Declaration of Independence—*all men are created equal*. It had turned out that not all people were equal in the United States. One large group of people was not even free.

Slavery had come to the land with the Spanish and English settlers. But they weren't the first to enslave people on the American continents. Some Native American nations had practiced slavery. It was an evil but common practice across the world. There were jobs that no one wanted to do and, in the days before machinery, slaves seemed an answer. So if you were on the losing side of a war, or were kidnapped by a rival tribe, or by a thief, you might end up as a slave.

In colonial times, there was slavery in both North and South. But slavery didn't make much sense in the North: farms were small and the farmer and his family could often handle the farm work themselves. After the Revolutionary War, slavery was outlawed in most northern states.

The situation was different in the South. From the earliest colonial days, the crops that grew well there—tobacco, cotton, rice, and sugar —demanded large numbers of fieldworkers. But there were few workers

Slaves were valuable, so most owners didn't abuse them. But some did. This man was put in chains and a "choker," and branded with a hot iron.

Every master of slaves is born a petty tyrant. They bring the judgment of heaven on a country. As nations cannot be rewarded or punished in the next world they must be in this. By an inevitable chain of causes and effects providence punishes national sins by national calamities.

—GEORGE MASON, AUTHOR OF THE VIRGINIA BILL OF RIGHTS, AND A SLAVE OWNER

There were machines in the South, as this illustration of cotton pressing shows. But what powered the machines?

I tremble for my country when I reflect that God is just. His justice cannot sleep forever.

—THOMAS JEFFERSON, VIRGINIA SLAVE OWNER

Three generations of a slave family dressed in their Sunday best for the photographer. Some families didn't get a chance to stay together; fathers, mothers, or children might be sold away.

to be had—until a Dutch ship arrived at Jamestown in 1619 with a boatload of Africans. At first the Africans were indentured servants. Then they became slaves. It solved an economic problem for the planters: slaves provided a cheap, easy source of labor. After the Revolutionary War, slave laws grew harsher and harsher in the South.

Slavery raised two issues besides economics. One was race. In the United States the slaves were all people of color: either Indians or blacks. And there was the issue of right and wrong. Some Northerners—and some Southerners—thought slavery morally wrong. Yet few of them were willing to do anything about it. The Southerners who opposed slavery did not speak up loudly and, as long as slavery stayed in the South, most Northerners were happy to forget about it.

But there were western lands coming into the country—and that was where problems developed. Southerners wanted slavery to expand. They wanted the new territories in the West to be slave territories. Northerners didn't.

Remember, most of the white people who didn't like slavery kept quiet. They didn't do anything about it. Was that wrong? Why didn't they speak out? Maybe because it wasn't easy to attack slavery. Those who did speak out weren't very popular. They were called *abolitionists* (ab-uh-LISH-un-ists) because they wanted to abolish, or end, slavery. Today we know that the abolitionists wanted to do the right thing. But if you want to understand history—to understand why things happened the way they did—you have to try to think as people did in the past. You have to put yourself in their times. Slavery had been around for a long time. People were just beginning to realize how terrible it was. Most Northerners didn't know any slaves, and many Southerners fooled themselves into thinking the slaves were happy.

Many people argued that if slavery were abolished it would wreck the South. Many businesses—in the North as well as the South—would be hurt. Change can be frightening. Some people said the abolitionists were troublemakers, and dangerous.

Some southern leaders said slavery had to go into the territories. Southern slave owners wanted to bring their slaves with them when they went visiting in the North. They said the whole nation needed to allow slavery.

A few sensible people tried to find ways to end slavery without destroying the southern economy. Ralph Waldo Emerson (who was a poet and writer) suggested that the government pay the slave owners for their slaves and then set them free.

That would have cost a whole lot less than going to war. But both sides rejected the idea. In the South, John Calhoun and other political leaders began to say that slavery was a wonderful, God-inspired system—good for slaves and good for the nation. Some people were foolish enough to believe them. Others, like Abraham Lincoln, saw clearly that slavery was evil and could only breed evil.

"We began," said Lincoln, "by declaring that all men are created equal; but now from that beginning we have run down to the other declaration, that for some men to enslave others is a 'sacred right of government.' These principles cannot stand together."

By 1860, there seemed to be no way around it. If the Union was to survive, and be true to its founding principles, there would have to be a war.

Solomon Northup, an escaped slave, wrote of slaves in the cotton fields: "They are not permitted to be a moment idle until it is too dark to see."

> **[Slavery] is the curse of heaven in the States where it prevails.**
> —GOUVERNEUR MORRIS, AUTHOR OF MUCH OF THE CONSTITUTION

A "sacred right of government"? What does that mean? It means a God-given right. Many in the South argued that slavery was part of God's plan. Explaining everyday events and problems in God's terms was not unusual in the 19th century.

1 Americans Fighting Americans

Each side thought the other side wasn't serious about it and they found, sadly, that war was much different than they thought it was going to be.

—ED BEARSS, HISTORIAN

SOUTHERN RIGHTS AT ALL HAZARDS says this scroll, illustrating a Charleston newspaper ad for artillery volunteers.

It was the worst war in American history. It was called the Civil War, or the War Between the States, and sometimes brother fought brother and father fought son. More than 620,000 Americans died. Cities were destroyed, farms burned, homes leveled, and, on one bloody day at a place called Antietam, more men were killed than on any other day in all our history. The total deaths were almost as many as in all of our other wars combined. If the same percentage of today's population were killed it would mean five million deaths.

A cartoon sneers at the leaders of the first six southern states to leave the Union. Each politician is shown as caring only about grabbing whatever he can for his own state.

The southern states gallop blindly off a cliff in this cartoon from a Northern paper.

It was the South against the North and, although the North won, neither side came out ahead. The South, which had once been prosperous, was in ruins. The North was left with huge war debts. And both North and South had the graves of fathers, sons, and husbands to weep over.

What was it all about? Why were Americans fighting Americans?

When the war began, people on both sides claimed they weren't fighting over slavery. But they were fooling themselves. Before the end of the war it was clear: slavery was the main issue. Most white Southerners wanted to keep slavery because they thought their way of life depended upon it.

You think slavery is right and ought to be extended, while we think it is wrong and ought to be restricted. That I suppose is the rub. It is certainly the only difference between us.

—ABRAHAM LINCOLN

15

Our new government is founded…upon the great truth that the negro is not equal to the white man.

—ALEXANDER STEPHENS, VICE PRESIDENT OF THE CONFEDERATE STATES OF AMERICA

To **secede** means to break away or withdraw from something, especially from a political organization or nation.

That big pussycat Abraham Lincoln vainly tries to stop Virginia from joining the other southern rats in their flight from the sinking Union ship.

Most Northerners thought slavery wrong and that, as Abraham Lincoln said, the nation could not exist half slave and half free.

There were other issues, too: the Southerners, who were also called Rebels, believed in *states' rights*. They thought any state should have the right to pull out of the United States (they usually called it the Union). They said it was tyranny to hold states in the Union against their wishes. They said they were doing the same thing that George Washington and John Adams and the other revolutionaries had done against King George: fighting for their freedom. But it was white freedom they were fighting for. They didn't want to consider the fact that they were tyrannizing black people.

What they did was form their own nation. Eleven southern states seceded from the Union. They created the Confederate States of America and elected a president and a congress. They said all they wished was to go peacefully from the Union.

The North wouldn't let them do it. Revolution is only right, said President Abraham Lincoln, "for a morally justified cause." But the South had no just cause. So, said Lincoln, secession was "simply a wicked exercise of physical power."

This was an important issue they were deciding. The American nation was still considered an experiment. Would a people's government survive? Lincoln said Americans needed to prove "that popular government is not an absurdity." Then he added, "We must settle this question now, whether in a free government the minority have the right to break up the government whenever they choose."

The Northerners, who were also called Yankees, or Federals, were willing to fight for the American form of government—for the Constitution, for the Union. They said that when the states joined the Union they agreed to uphold the Constitution and they couldn't just pull out anytime they wanted. If that were allowed, soon there would be no Union at all.

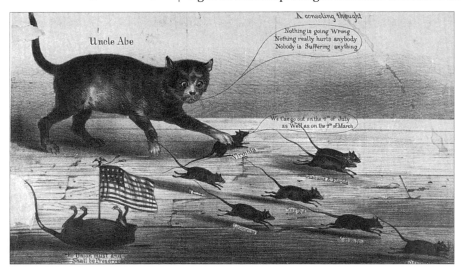

2 The War Begins

The tea has been thrown overboard, the revolution of 1860 has been initiated.
—THE *CHARLESTOWN [S.C.] MERCURY* ON FORT SUMTER

Siege guns, like this 15-inch cannon (15 inches was the diameter of the gun's bore—the inside of the barrel), battered Fort Sumter into submission after 34 hours.

Most people, North and South, thought the war would last a few months. The Southerners liked the idea of soldiering. It seemed adventurous and heroic. Many Southern leaders were graduates of the military academy at West Point. Besides, they were sure the Yankees were all cowards. Just wait until they met on the battlefield, they'd scare the wits out of those Yankees. Or so they boasted to their wives and girlfriends as they marched off in their handsome gray uniforms.

The Northern men were just as confident. One big battle, they said, and the war would be over. They knew the North had many advantages: more men, more industry, more money. But that wasn't what made them confident. They were sure the Southern soldiers were lazy. Why, without their slaves they wouldn't be able to do a thing. They'd run for the hills at the first shots, or so the Northerners boasted to their wives and girlfriends as they marched off in their handsome blue uniforms.

Father and I were husking out... corn... when William Corry came across the field. He was excited and said, "Jonathan, the Rebels have fired upon Fort Sumter." Father got white, and couldn't say a word.
—THEODORE F. UPSON, LIMA, INDIANA

"A perfect sheet of flame flashed out, a deafening roar, a rumbling, deadening sound, and the war was on," wrote a Charleston teenager of the bombing of Fort Sumter.

17

The ladies of Charleston watched the bombing of Ft. Sumter. "I had a splendid view," wrote Emma Holmes. "With the telescope I saw the shots as they struck the fort, and the masonry crumbling."

The United States has usually prepared for its wars after getting into them. Never was this more true than in the Civil War. The country was less ready for what proved to be its biggest war than for any other war in its history.

—JAMES McPHERSON, *BATTLE CRY OF FREEDOM*

It didn't quite turn out the way they expected. Men on both sides were brave, very brave, and willing to die for their beliefs. A generation of men would die. But no one realized that in the summer of 1861.

The war began when Southern guns fired on U.S. troops stationed at a small island fort in the harbor at Charleston, South Carolina. That fort—Fort Sumter—was a United States government fort, and those shots announced that South Carolina was serious about leaving the United States. The Southerners meant to capture the fort, and they did. A few South Carolinians were dismayed, but many came to the Charleston waterfront in party clothes and cheered when the cannons blasted. They were rebels—eager to battle what had been their own government—and they were excited by what they were doing. Some of the young men who applauded the firing were afraid the war would be over before they got a chance to fight. One political leader said he would use his handkerchief to wipe up all the Southern blood that would be spilled—and people believed him.

The first big battle was fought at a place called Manassas, not far from the city of Washington. It was fought near a muddy stream known as Bull Run. So some people call it the battle of Bull Run, and others the battle of Manassas.

Manassas was a logical place to have a battle. It was a railroad junction: the place where two railroad lines met. This would be the first war where modern transportation was used. Again and again, the railroads would make a difference. They would help decide this battle of Manassas.

The Northern generals thought they would take Manassas and then march south, to the Confederate capital at Richmond, Virginia.

Here is what Bruce Catton (one of

When General P. G. T. Beauregard (above) led the Confederate attack on Fort Sumter, the fort's commander was Major Robert Anderson, a former slave owner from Kentucky who stayed loyal to the U.S.A.

the best of many writers on the Civil War) said about the battle:

> *There is nothing in American military history quite like the story of Bull Run. It was the momentous fight of the amateurs, the battle where everything went wrong, the great day of awakening for the whole nation, North and South together. It...ended the rosy time in which men could dream that the war would be short, glorious and bloodless. After Bull Run the nation got down to business.*

When that July day began, in 1861, war seemed a bit like a show. And hundreds of Washingtonians didn't want to miss that show. (After all, they couldn't watch it on TV.) They decided to go to Manassas with their picnic baskets, settle down near Bull Run stream, and watch the fighting. They came on horseback and by carriage and wagon and they spread out in the fields and listened to the guns and watched a smoky

We...fired a volley, and saw the Rebels running...The boys were saying..."We'll hang Jeff Davis to a sour apple tree." "They are running." "The war is over."
—PRIVATE JAMES TINKHAM, MASSACHUSETTS VOLUNTEER

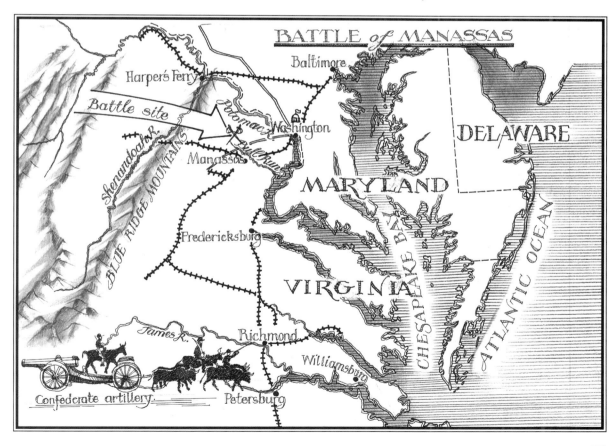

19

At the start of the war, volunteers, like these Confederate soldiers, enlisted eagerly. "We were all afraid it would be over and we not in the fight," said 21-year-old Sam Watkins, whose family owned no slaves.

haze form above the noisy cannons.

But they didn't see what they expected. It wasn't a picture-book battle; it was real, and disorderly. They were all untrained beginners out there—soldiers and officers—and they didn't know what they were doing.

The officers were trying to fight war as Napoleon had, for Napoleon was considered the military genius of the age. But Napoleon had a trained army able to follow complicated military orders. And guns had changed since Napoleon's day—they were now more deadly and accurate. It took time for the officers to realize that the old tactics needed to be thrown out. Besides, American soldiers weren't like Europeans. The American soldier was an independent kind of fellow; he didn't take orders well. But he could fight like fury, as everyone soon found out.

Both sides fought all day. They fought hard, and even though there was much confusion, there was also much bravery. One of the bravest of the fighters was a Confederate general, T. J. Jackson. "There is Jackson standing like a stone wall," shouted a Southern officer when he saw Jackson and his men holding firm against the enemy. Jackson would, from that moment, always be known as Stonewall Jackson.

Stonewall Jackson

The battle raged, over meadows and wooded hills and on the steep banks of Bull Run Creek. It was a hot, very hot, humid summer day. Most of the men wore heavy wool uniforms, as was customary then. Those uniforms were anything but "uniform." Some were official Southern gray or Northern blue. But many men wore the uniforms of their state militias (some southern militias had blue uniforms). Some men had borrowed old Revolutionary War uniforms. Some wore fancy wide pants and colorful

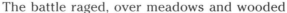

The Ladies of Savannah

A letter to his parents from Charles C. Jones, Jr., mayor of Savannah, Georgia

My very Dear Father and Mother:
The ladies of Savannah are not idle. They are daily engaged...in the preparation of cartridges both for muskets and cannon. Thousands have been already made by them, and the labor is just begun. Others are cutting out and sewing flannel shirts. Others still are making bandages and preparing lint [lint was padding used to cover wounds].

sashes copied from French Zouave soldiers. Some just wore their regular clothes. It was confusing, and more than a few soldiers were killed by bullets shot from their own side.

By afternoon everyone was exhausted, bodies littered the ground, the earth was bloody and beginning to smell, and neither side seemed to be winning. Then fresh Southern troops arrived by train. That made the difference. It gave new energy to the rebels. General Jackson—Stonewall—told them to "yell like furies." They did. They attacked with bloodcurdling shouts; they called it the "rebel yell." And that was too much for the Yankees. They dropped their guns and fled. Some couldn't run fast enough.

The South won that battle of Bull Run. The Northern soldiers, who had planned to fight on to Richmond, now went the other way, back to Washington. They hadn't expected to do that. The congressmen and the citizens who had come to watch the battle hadn't expected it either.

Troops and civilians were together on the road back to Washington when a stray shell exploded and upset a wagon. That blocked a bridge, and wounded soldiers, cavalry, frightened troops, and families in their carriages were all stuck. Someone shouted that the Southern cavalry was attacking. (It wasn't really.)

That was like yelling

Tearful farewells to the troops at the railroad station.

A Unionist wrote of the untidy Confederate troops, "Were these dirty, lank, ugly specimens...the men that had...driven back again and again our splendid legions with their fine discipline?"

They plunged through Bull Run wherever they came to it, regardless of fords or bridges, and there many drowned....We found...along the road, parasols and dainty shawls lost in their flight by the fair, fair ones who had seats in most of the carriages.

LT. COL. W. W. BLACKFORD,
1ST CAVALRY, VIRGINIA

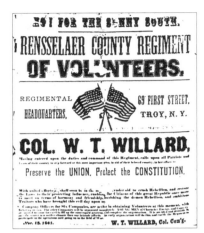

HO! FOR THE SUNNY SOUTH.

RENSSELAER COUNTY REGIMENT
OF VOLUNTEERS.

REGIMENTAL
HEADQUARTERS,
69 FIRST STREET,
TROY, N.Y.

COL. W. T. WILLARD,

Preserve the UNION, Protect the CONSTITUTION.

W. T. WILLARD, Col. Com'f.

"Ho! for the Sunny South," says this Union recruiting poster. The war would be just a summer vacation.

One Union general said of Bull Run: "The retreat soon became a rout, and this degenerated into a panic."

"Fire!" in a crowded theater. It started a panic—a shoving, pushing, screaming panic.

It didn't take long for people to realize that war is no picnic. Although, even then, no one dreamed that the war would be as long, hard, and bloody as it turned out to be.

How did that terrible war actually come about? What caused all the anger? How did good people in the North and good people in the South come to hate each other?

Could slavery have been ended without war? Could a civil war happen again? What can a nation learn from its history?

Flash back in time with me. Perhaps we can find some answers. We'll start with the stories of two men and two women—four Americans who were caught in their times and willing to fight for their beliefs.

Each of the women was named Harriet. Each was small in size (yet a giant in strength and determination). Each became an American heroine.

The two men were born in Kentucky. Both were men of integrity—which means they could be trusted to do what they thought was right. But what was right for one seemed wrong to the other. That different way of looking at things was what the war was all about. The two men were enemies. Read on, you'll find their stories absorbing.

3 Harriet and *Uncle Tom*

Harriet Beecher Stowe and her most famous brother, Henry Ward Beecher; do you see a family resemblance?

Harriet Beecher grew up in Connecticut in a houseful of children: seven boys and four girls. Her father, Lyman Beecher, was a Congregational minister, known throughout New England for his fine sermons. The Congregationalists were descendants of the Puritans: serious and moral in their religion. All Lyman Beecher's boys became ministers. One of them, Henry Ward Beecher, was said to be the greatest preacher of his day. Thousands of people packed his church when he spoke. Another son, Edward, was a college president.

Catharine and Isabella, two Beecher daughters, were pioneers in the fight for women's rights.

Harriet was the smallest of Lyman and Roxana Beecher's children. She never grew to be more than five feet tall. But size has nothing to do with ability: she was the sister everyone depended upon. She had a way of getting things done. She became the most famous Beecher of all. In fact, Harriet Beecher, who married Calvin Stowe and then was Harriet Beecher Stowe, was the most famous American woman of her day. And all because of a book she wrote, a book that changed history.

You can see that the Beecher family wasn't ordinary. The Reverend Lyman Beecher played the violin, sang hymns, and was known to dance around the house in his stocking feet. He was very religious; he also had a good sense of humor. He liked to ask his children questions and make them think. If a child couldn't beat him in an argument, Lyman would give out clues to help sharpen the debate.

Finding ideas was easy for Lyman Beecher; it was his watch that he

The Beecher home still stands in Litchfield, Connecticut.

When she was grown up, Harriet Beecher Stowe described herself as "a little bit of a woman… about as thin and dry as a pinch of snuff."

At the age of 12, Harriet read her favorite book, *Ivanhoe*, seven times in *one* month.

The Reverend Lyman Beecher (front row center) with nine of his eleven children. Harriet is on the far right with Henry behind her; Isabella is on the far left with Catharine next to her.

Anguish means sadness and pain.

could never find. He was always losing things: one day it was his sermon, the next day his hat. When he didn't lose things he gave them away. Like his new coat, which he gave to a poor man who didn't have one.

Once, on a Sunday, the Reverend Beecher exchanged churches with another minister. It was just for one day and seemed a good idea. But, while the other minister was preaching the Beecher dog entered the church. Now the dog often did that, but this time he saw a strange man on the pulpit. So he walked right up there and started barking. That was bad enough, but then one of the Beecher children began giggling. And then another one. Soon the whole family was shaking with laughter. Finally they all had to march out of church—children, dog, and mother too.

When Harriet was 13, her sister Catharine, who was 24, opened a school for girls. Harriet went as a student, but then she began teaching as well. She discovered what every teacher knows: teaching is a good way to learn. Harriet was also beginning to write, and to care about words. Then Lyman Beecher moved to Cincinnati, Ohio, where he became head of a college to train ministers. He wanted his family with him, so Catharine and Harriet became Midwesterners.

In New England, slavery had seemed far away. Now it was close by. Ohio was a free state: there was no slavery there. But Kentucky, just across the Ohio River, was a slave state. Harriet stood on the banks of the river and watched boats filled with slaves in chains who were being shipped south to be sold at slave markets. One day she saw a baby pulled from its chained mother's arms. She saw a look of anguish on the mother's face. She never forgot that look.

Once, Harriet was invited to Kentucky to visit a friend who lived on a plantation and owned slaves. The friend and her family were kind people, and she saw slavery at its best. But when Harriet and her friend rode horses to a neighboring plantation, they saw a cruel overseer abusing blacks. Harriet remembered the kindness and the cruelty, for she was becoming a writer and so she observed and remembered.

She wrote stories and essays and poems. Then, when she married Calvin Stowe and had babies, there never seemed to be enough money.

Calvin was a teacher in her father's college, and not paid a high salary. So Harriet wrote stories to earn money.

Calvin got a job teaching at Bowdoin College in Maine. By this time Harriet had learned a lot about slavery, and it made her very angry. Her brother Edward's wife said to her, "If I could write as you do I would write something to make this whole nation feel what an accursed thing slavery is." And that was just what Harriet Beecher Stowe did. She wrote *Uncle Tom's Cabin*.

It was the most important American book written in the 19th century. It may be the most influential book ever written in America. Its chapters were first printed in a newspaper. Within a week of its publication as a book (in 1852), 10,000 copies had been sold—and *Uncle Tom's Cabin* was just getting started. Before the Civil War even began, two million copies were bought in the United States; and it was translated into many languages and sold around the world. Historian James

Harriet (left) wrote at the kitchen table with her children running in and out. She dreamed of a place to work without "the constant falling of soot and coal dust on everything in the room."

"Harriet Stowe had made a black man her hero, and she took his race seriously, and no American writer had done that before," wrote the historian David McCullough. Below, from *Uncle Tom's Cabin*, Topsy and Uncle Tom himself.

· UNCLE · TOM ·
· UNCLE · TOM'S · CABIN ·

135,000 SETS, 270,000 VOLUMES SOLD.

UNCLE TOM'S CABIN

FOR SALE HERE.

AN EDITION FOR THE MILLION, COMPLETE IN 1 Vol. PRICE 37 1-2 CENTS.
"   "   IN GERMAN, IN 1 Vol. PRICE 50 CENTS.
"   "   IN 2 Vols. CLOTH, 6 PLATES, PRICE $1.50.
SUPERB ILLUSTRATED EDITION, IN 1 Vol. WITH 153 ENGRAVINGS,
PRICES FROM $2.50 TO $5.00.

The Greatest Book of the Age.

"TOPSY"
· UNCLE · TOM'S · CABIN ·

In Maryland, Samuel Green, a free black, was sentenced to 10 years in prison for having copies of *Uncle Tom's Cabin* and some other antislavery works.

Calling someone an "Uncle Tom" today is an insult. It means a black person who behaves humbly or submissively toward white people. The Uncle Tom in Harriet Beecher Stowe's book is a heroic figure.

In an early illustration from *Uncle Tom's Cabin*, slaves rejoice as they are freed by the owner of their plantation.

McPherson says it was "the best seller of all time in proportion to population." In England, Lord Palmerston read it three times. (A decade later, he was Britain's prime minister and had to decide whether to get involved in America's civil war or remain neutral.)

It is a very exciting book. Some critics say it is not great literature, but you can judge that for yourself. *Uncle Tom's Cabin* is good reading, and not difficult, except that it is full of dialect, the everyday talk of blacks in the old South. If you can handle the dialect you will like the book—a lot. Anyone who reads *Uncle Tom's Cabin* and doesn't cry at the end has a hard heart. It was the first American novel to make real people of blacks, and it made people care. Harriet Beecher Stowe tried to be fair when she wrote the book. She made the horrible overseer, Simon Legree, a Northerner. Legree is the villain in the story. Uncle Tom, a saintly black man, is strong and heroic, the finest person in the book. Two black men who beat him are evil. Some white plantation owners are good people. Harriet was trying to show that color has nothing to do with whether a person is good or bad. What she showed very well was that the system of slavery was evil and that even good people did evil things when they were part of that system.

Uncle Tom's Cabin changed people's ideas about slavery. It made people in the North angry. It made them willing to fight a war to end slavery. In much of the South it was against the law to buy or sell the book. When President Abraham Lincoln met Harriet Beecher Stowe during the Civil War he said to her, "So this is the little lady who wrote the book that made this great war."

4 Harriet, Also Known as Moses

Harriet Tubman's faith was strong. She said, "I always told God, I'm going to hold steady on you, and you've got to see me through."

Harriet Beecher Stowe didn't make the Civil War, but she did help make people want to fight that war. She wasn't the only one who did that. There were others. One was Harriet Tubman, also known as Moses, who was born a slave.

There was none of the spirit of an enslaved person in Harriet Tubman. She was a fighter: tough, brave, and brilliant. Like the other Harriet she was tiny—just five feet tall—but this Harriet was stronger than most men. She could lift great weights, withstand cold and heat, chop down big trees, and go without food when necessary. She had been trained, in childhood, to take abuse. That was part of what it meant to be a slave. She put that training to good use.

Harriet never knew for sure when she was born; few people recorded slaves' birthdays. But it was in Maryland, on what is known as the Eastern Shore, in about 1820. (The Eastern Shore is the peninsula that looks a bit like a fist, with a finger pointing to Norfolk, Virginia, and its wrist near Wilmington, Delaware.)

Her owner considered her a problem child. He sent her off to work when she was six. He got the wages, she did the work. She was sent to dust and sweep and tend a baby. But she didn't know how to dust or sweep. She had grown up in a slave cabin with a dirt floor and no furniture—just old blankets to sleep on. The woman she worked for didn't think about that. When Harriet didn't dust well she beat her.

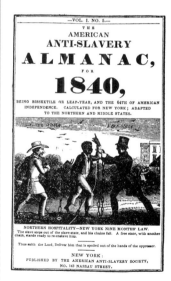

The cover of this abolitionist almanac protested a New York law that said escaped slaves had to live in the state for nine months before claiming residency—which gave the slave catchers time to find them.

Hard. When the baby cried, Harriet was supposed to rock her and make her stop crying. Sometimes little Harriet fell asleep. Her mistress beat her. Hard.

Harriet ran away. But she didn't know where to go, so she hid in a pig pen. Living with pigs may have been easier than living with cruel humans, but she didn't have anything to eat except the potato peelings the pigs ate. Soon she was starving and had to go back. This time her mistress beat her so hard she carried scars for the rest of

Despite the rather romantic view of escape suggested at right, most slaves ran away alone—it was simpler and safer. This ordinary-looking house, in Canisteo, New York, was a station on the Underground Railroad to Canada.

There was one of two things I had a right to, liberty, or death; if I could not have one, I would have the other; for no man should take me alive; I should fight for my liberty as long as my strength lasted, and when the time came for me to go, the Lord would let them take me.

—HARRIET TUBMAN

her life. Now Harriet was too sick to work at all, so she was sent back home. Her mother and father nursed her back to health. The master sent her out to work again: this time to a house where the woman was a weaver and the man a hunter. They were just as cruel as the first people. Harriet was supposed to help with the weaving, but she hated the work and was no good at it. She was sent to check the hunter's traps. She had to wade through cold water. One day she said she was too sick to go. The hunter didn't believe her. Harriet had the measles and bronchitis. After going in the cold water she almost died. She was sent back home again. Again her parents cared for her, but her throat had been damaged. After that she always had a low, husky voice.

She was happy to be back home with her parents and her ten brothers and sisters. The slave owner could see that she was no good to hire out, so he sent her to work in the fields. Even though she was still very young and very small, she did hard work and she liked being outdoors. Of course, slaves couldn't go to school, so no one taught Harriet to read and write, but she learned to listen and to remember, and she soon had an unusual memory. She listened to the blacks who whispered about freedom. She learned that a few slaves were freed by their masters. She learned that others ran away north, and found freedom. She learned that if a slave tried to escape and was caught, he would be whipped, branded, and sold. He would be sold south, far south, to cotton plantations where life was even harder for blacks than it was in Maryland and there was little chance of escape.

Some blacks were sold south just because the master needed money (and all lived in fear of that). Two of Harriet's sisters were sold, and her parents grieved as parents grieve for dead children.

Harriet learned that some people—black and white—helped escaping blacks. They were part of something called the *Underground Railroad*. It wasn't a real railroad, although Harriet thought it was when she first heard of it. The Underground Railroad was a way to get north. It was a series of places where blacks would find help. The places—houses, barns, and boats—were called *stations*. People who traveled the route were called *passengers*. People who led them were *conductors*.

But most of what Harriet heard from the whispering was guesses. No one who escaped ever came back. No one really knew what happened

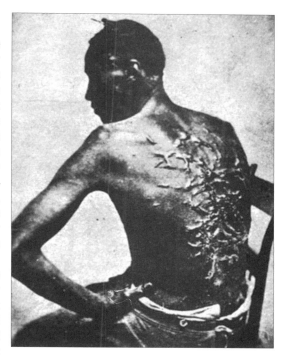

On Solomon Northup's plantation, "it was rarely that a day passed by without one or more whippings....One hundred [lashes] is called severe; it is the punishment inflicted for ...standing idle in the field."

Them days was hell. Babies was snatched from their mother's breast and sold. Childrens was separated from sisters and brothers and never saw each other again. Course they cried. You think they didn't cry when they was sold like cattle?

—WORDS OF WOMEN SLAVES, QUOTED IN GEOFFREY WARD, *THE CIVIL WAR*

Escaped slaves lived with the fear that slave catchers would find them. In 1856 they caught Margaret Garner and her four children in Cincinnati. Garner decided she would rather die than be a slave again. She killed one of her children and would have killed the others and herself, if she had not been stopped.

Some nights I dream about the old slave times and I hear them crying and praying, "Oh master, pray! Oh, master, mercy!" when they are being whipped, and I wake up crying. I sit here in this room and can see it plain as day, the hard work, the plantation, the whippings, the misery.
—BEN BROWN, FORMER SLAVE, INTERVIEWED AT AGE 99 IN 1936

to them. Harriet would be different: she would escape and she would come back. That was in the future; she still had some growing up to do.

One day Harriet was in a store. An overseer was there. Suddenly he yelled at a slave who was running. Then he yelled at Harriet to stop the slave. Harriet didn't move. The overseer threw a lead weight at the running slave. The weight hit Harriet, right in her forehead. She passed out. For months she lay unconscious. Everyone thought she would die. For the rest of her life she had fainting spells and times when she would just fall asleep and no one could wake her.

Her master decided to sell her, but no one would buy a slave who was always falling asleep. (And Harriet made sure she fell asleep whenever a buyer was around.) She was sent to work with her father, who was one of the most trusted slaves on the plantation. He never told a lie. He was in charge of a woodcutting operation. Harriet soon learned to cut trees as easily as the strongest of men.

Her father taught her the ways of the woods. He taught her to walk softly, as the Native Americans did. He showed her the plants she could eat, and the ones that were poisonous.

Then she met a white woman who said she could help her. Harriet guessed that the woman was part of the Underground Railroad. She guessed right.

I can't tell the whole story of Harriet Tubman. It takes a book to do that. I will tell you that after she escaped north, to freedom, she didn't feel free without her family and friends. So she went back and got them. She got all of her brothers and sisters and her parents—except for the two sisters who were sold south. It took many trips, and it wasn't easy.

She didn't stop with just her family. She became the most famous conductor on the Underground Railroad. She is said to have led 300 blacks to freedom.

She made Northerners think differently about slavery. Southerners, like Senator John Calhoun, were saying that slavery was a good thing. They said slaves were well treated and loved their masters and mistresses. But people who are well treated don't risk everything—sometimes even their lives—to run away. The blacks who escaped on the Underground Railroad told of children being taken from their parents; they showed scars from whippings; they told of abuses that made

"I saw...a brawny, famished looking, black man...bearing over his head, something in the form of an arch...three feet above his hair, beneath... which were suspended the bells [an antirunaway device]."

many Northerners change their minds about slavery.

Try to imagine that you are going north on the Underground Railroad. You'll travel at night, the darker the better, so you can escape the slave catchers. Follow the North Star and you'll be going in the right direction. Keep off the roads as much as possible—the woods are safer. If wild animals and snakes bother you, too bad: no screaming or loud talking allowed. If you hear dogs you'd better make for a river, even if it's freezing cold. Dogs can't follow your scent in the water. If you see strangers you'll have to think and act quickly.

Harriet Tubman was a quick thinker. Once, she and some slaves were about to board a train heading north. She saw some slave catchers. So, quickly, she had everyone get on a train going south. No one suspected slaves heading south.

She was daring and ingenious and soon there was a huge reward for her capture. Often she wore disguises: sometimes she

She Wd Whistle

Martha C. Wright, an abolitionist, described one of Harriet Tubman's rescues in 1860.

We have been expending our sympathies, as well as congratulations, on seven newly arrived slaves that Harriet Tubman has just pioneered safely from the southern part of Maryland. One woman carried a baby all the way and bro't two other chld'n that Harriet and the men helped along. They bro't a piece of old comfort and a blanket, in a basket with a little kindling, a little bread for the baby with some laudanum [opium] to keep it from crying during the day. They walked all night carrying the little ones, and spread the old comfort on the frozen ground, in some dense thicket where they all hid, while Harriet went out foraging, and sometimes cd not get back till dark, fearing she wd be followed. Then, if they had crept further in, and she couldn't find them, she wd whistle, or sing certain hymns and they wd answer.

Harriet Tubman in old age. She lived to be over 90.

Breaking the Law: A Discussion of Ethics

Today we think of the people who ran the Underground Railroad as heroes. But they were lawbreakers. The law said that runaway slaves were to be returned to their owners. One prominent northern senator, William Henry Seward, said, "There is a higher law than the Constitution." He was talking about God's law. He thought slavery broke God's law and should therefore be opposed. His speech was widely criticized. It was called "revolutionary." Southern senators could find passages in the Bible that dealt with slavery. They said slavery was part of God's plan.

What do you do if you think a law is evil? Do you break it? Do you search for God's law? Where would you find it? In the Bible?

Whose bible? The Koran? The Hebrew Bible? The New Testa-

The abolitionists were the first reformers to use mass-media methods—like this ABC book— to spread their message.

ment? The Bhagavad Gita? Whose interpretation of the holy book do you choose?

What happens if you break the law? You know: you go to jail.

All of this is a philosophical discussion. (Philosophy is the study of ideas.) If you believe in a government of laws—and most Americans do—then breaking a law is a very serious and difficult thing to do.

In the 19th century, people in the North and South who hated slavery were faced with some of these questions. Yes, there were people in the South who hated slavery. Some of them taught slaves to read. They broke the law. Some helped slaves escape. They broke the law.

I have an answer to this

William Henry Seward

THE

ANTI-SLAVERY

ALPHABET.

"IN THE MORNING SOW THY SEED."

PHILADELPHIA.
PRINTED FOR THE ANTI-SLAVERY FAIR
1847.

Merrihew & Thompson, Printers.

dilemma of the evil law, but before I tell you my thoughts, come up with answers of your own. Then see if we agree. You are entitled to your own opinion.

Okay, here is my opinion.

dressed as a man, sometimes she pretended to be a very old woman. Once she saw her old master coming toward her. She knew he would recognize her. She was carrying some live chickens. So she let them go and chased after them. The man laughed and yelled, "Go get them, Granny." He had no idea that Granny was Harriet Tubman, the most wanted runaway slave in the nation.

Harriet never stopped fighting for what she thought was right, never for a minute. When the Civil War began she was asked to help the Union army. She did.

She went to South Carolina, where again she helped blacks escape; she also went behind the lines as a scout and spy. She went to Virginia and worked as a nurse in an army hospital at Fortress Monroe. She was never paid for her work, but lived by her wits, buying and selling

Government under law is the only reasonable form of government in our complex world. So it is important that citizens respect the law. However, laws are made by human beings, and sometimes people make mistakes. Sometimes laws are unjust. The best way to handle an unjust law is to work to get it changed. Write your senator or representative in Congress. Even better, if you are old enough, get elected to Congress.

But suppose the unjust law doesn't get changed. Then, sometimes, for some people, the best course is to break the law.

That's a big decision. Lawbreakers must be prepared to pay a price. They must be prepared to go to jail. The writer Henry David Thoreau was willing to go to jail for his beliefs (see Book 5 of *A History of US* for a chapter about Thoreau). Some of the abolitionists went to jail for helping slaves. That is one of the reasons they were heroic.

Read the Declaration of Independence again, and you will see how the men who founded this nation felt about resistance to evil laws.

One group of people has no right at all to break the law. They are government officials who have been elected or appointed to uphold the law. They must resign their offices if they feel they cannot enforce an unjust law.

Being an intelligent, ethical citizen isn't always easy.

The problem of the unjust law will be around as long as there are people and governments. One hundred years after the Civil War, a Christian minister named Martin Luther King, Jr., wrote a letter from jail in Birmingham, Alabama. The laws in Alabama said that blacks could not sit in "white" restaurants or sit where they wished on public buses.

King thought those laws unjust; he broke them and went to jail. In jail he wrote, "An individual who breaks a law that conscience tells him is unjust, and willingly accepts the penalty by staying in jail...is in reality expressing the highest regard for the law." Do you agree? Or disagree? Why? (See what you can learn about a man named Mohandas Gandhi. What is *civil disobedience*?)

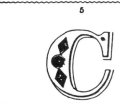

C is the Cotton-field, to which
This injured brother's driven,
When, as the white man's *slave*, he toils
From early morn till even.

D is the Driver, cold and stern,
Who follows, whip in hand,
To punish those who dare to rest,
Or disobey command.

A page from The Anti-Slavery Alphabet.

vegetables, baking pies, and brewing root beer in order to eat.

After the war she moved to Auburn, New York, married a war veteran, cared for her parents, and opened a home for elderly blacks. She was often asked to speak of her experiences, and she did. She spoke out for equal rights for women. Though she had never been lucky enough to go to school, she knew the importance of learning. She worked to get schools for the newly freed blacks. The queen of England, Queen Victoria, heard of Harriet Tubman and wrote her a letter inviting her to come to England for a visit. Harriet didn't have the money to go.

Maybe it didn't matter. She became more famous than most of the people who curtsied to the queen. Few people in American history had the courage and determination of Harriet Tubman. Now do you know why she was called Moses?

Harriet Tubman's great-great-niece Mariline Wilkins says that Tubman knew about remedies and helped cure soldiers during the Civil War. The family believes she "discovered penicillin." Penicillin is found in mold, and Harriet Tubman used to take mold, put it in a jar, and add lemon juice, honey, and whiskey. The mixture didn't taste good, says Wilkins, but "it was good for colds."

5 Abraham Lincoln

The baby who was born in this Kentucky log cabin in 1809 became the first U.S. president to be born outside the original 13 states.

Abraham Lincoln's brother Jacob Lincoln fought with George Washington during the Revolutionary War. Abraham was a captain in the Virginia Militia.

Now don't get confused. This Abraham Lincoln of 1776 did not become president. It was his grandson who did. Grandfather Abraham never even knew his grandson. You see, he was killed by an Indian bullet. That was after he moved to Kentucky. He went there because his good friend Daniel Boone told him of the fine land in Kentucky. But it was Indian land, and the Native Americans would not give it up easily. If the English-speaking settlers were to have it they would have to fight for it.

Abraham's son, Thomas, who was only eight, saw his father killed. When Thomas grew up he told his son, young Abraham, stories of his grandfather and Daniel Boone and Indians.

They were poor folk, those Lincolns. They didn't have much except stories and hope. That was enough to keep them on the move, seeking good land and good fortune. It never seemed to happen.

Thomas did have good luck with the two women he chose to marry. Nancy Hanks, who was first, was shy and quiet and loved God and the Bible. Her son, Abraham, was born on a cold Sunday in February of 1809 in a cabin made of logs with a floor of hard-packed dirt. The cabin had one small window, a fireplace, and a door that swung on hinges made of animal hide. It was no different from the slave cabins on many Virginia plantations.

When Abraham was born, his 10-year-old cousin, Dennis Hanks,

Lincoln's cousin Dennis Hanks was 10 years old when Abraham was born. He married Abe's stepsister Elizabeth.

34

came and spent the night. Some years later, Dennis was asked about the baby. "Well now," said Dennis, "he looked jist like any other baby at fust, like a red cherry-pulp squeezed dry in wrinkles. An' he didn't improve none as he growed older. Abe was never much fur looks." When Dennis held him, baby Abraham howled. "He'll never come to much," said Dennis, and handed the baby back to his mother.

The baby grew—long and skinny, with big feet and dark hair. As he grew, he heard stories told aloud. From his mother he heard of Adam and Eve and Noah and of others in the Bible. From his father, who was a carpenter and a farmer and a good storyteller, he heard tales of the mountain folks.

Abe had a sister, Sarah, who was older by two years. Sometimes they went to school. It was called a "blab" school. There were different ages in the log cabin, which was just one room with a door and no window. Everyone said his or her own lessons—out loud—all at the same time, so the noise was like one big blab.

The Lincolns were members of the South Fork Baptist Church in Kentucky. Kentucky was a slave state, but some of the Baptists hated slavery and thought it wrong. The Lincolns felt that way. Other members of the church approved of slavery. The fights between those who were for slavery and those who were against got so bad that finally the church had to close its doors.

Soon after that, Nancy and Thomas and their two children and Dennis Hanks all moved across the river to the free state of Indiana. Indiana's constitution said, *All men are born equally free and independent*, and *the holding of any part of the human creation in slavery…[is] tyranny.* Land in Indiana was two dollars an acre.

Thomas went first. He hacked a trail through the deep woods and claimed land that looked good to him near Little Pigeon Creek. Then he took some logs and built a three-sided shed. The family spent its first winter in that hut with a blazing fire on the fourth side to frighten off wolves and mountain lions and to keep them warm. Even for a pioneer family, it was a primitive way to live. There were blizzards and

Abraham was a tall, strong boy and very good at splitting wood—though this picture was done as campaign propaganda years after Lincoln had stopped having to do such jobs.

Tyranny is the unjust or cruel use of power. Thomas Jefferson said, "I have sworn eternal hostility to every form of tyranny over the mind of man."

35

Wilderness Lingo

His folks...called themselves "pore" people. A man learned in books was "eddicated." What was certain was "sartin."...joints were "j'ints"; fruit "spiled" instead of spoiling; in corn-planting time they "drapped" the seeds. They went on errands and "brung" things back. Their dogs "follered" the coons. Flannel was "flannen,"...a chimney a "chimbly," a shadow a "shadder," mosquitoes plain "skeeters." They "gethered" crops. A creek was a "crick," a cover a "kiver."...They made their own words. Those who spoke otherwise didn't belong, were "puttin' on." This was their wilderness lingo; it had gnarled bones and gaunt hours of their lives in it.

—CARL SANDBURG, ABE LINCOLN GROWS UP

Sarah Lincoln in old age. After Abraham moved away from his family he didn't see them much, but he loved Sarah and helped provide for her welfare.

soaking rains, and when the wind whipped at the fire the smoke drove them from the hut.

Dennis said, "I don't see how the women folks lived through it. Boys are half wild anyhow, and me 'n Abe had a bully good time."

Later, Abraham wrote, "We reached our new home about the time the state came into the Union. It was a wild region, with many bears and other wild animals still in the woods. There I grew up."

Hunting was easy. Besides the woodland animals—raccoons, squirrels, opossum, deer, bear, wildcats—flocks of ducks, geese, and wild parakeets flew overhead and passenger pigeons were so thick they made the sky dark. Young Abe took his father's rifle, aimed through a chink in their new log cabin, and shot a wild turkey. He found he didn't like killing animals, and never shot anything larger than that turkey.

When Abe was nine his mother died of the milk sickness, an illness that poisoned cattle first and then people who drank milk from infected cows. Nancy Hanks Lincoln had been a deeply religious woman and now there was no minister to bury her. That made the family's grief harder to bear. Young Abraham wrote a letter to Kentucky, to the minister they had known there, and the following spring the Reverend David Elkin came to Pigeon Creek and said prayers over her grave. After the death of their mother, Abraham and his sister ached with loneliness, and they turned to each other and became especially good friends.

A man on the frontier with a family needs a woman, and so, within a year, Thomas Lincoln married Sarah Bush Johnston, who had three children of her own.

Sarah was a fine, industrious woman who came into the wilderness and tried to make something of the Lincoln clan and did a good job of it. She got Thomas to put a floor in the log cabin and to make good beds and other improvements. She took Dennis, Abe, and Sarah to the spring and soaped and scrubbed. Then she mended their old garments and made new clothes for them. She did more than that. She gave them love and encouraged them to learn. Abraham didn't need much encouraging. More than anything else he seemed to love to read. But books were scarce in Indiana—so he read the same ones over and over. He read Aesop's *Fables*, *Robinson Crusoe*, the *Arabian Nights*, the family Bible, and anything else he could find.

He borrowed *The Life of George Washington* and put it in bed beside him. One night snow came through the chinks in the roof and wet the book through. At first Abraham didn't know what to do, but his father had taught him always to be honest. So he went to Josiah Crawford, the man who owned the book, told him what had happened, did some work for him, and got the book as a gift.

All that book reading seemed strange to his neighbors, who couldn't read and didn't see the sense of it. One neighbor said, "[Abe Lincoln] worked for me....[he] was always reading and thinking. I used to get mad at him....I say Abe was awful lazy. He would laugh and talk and crack jokes and tell stories all the time....Lincoln said to me one day that his father taught him to work but never learned him to love it."

He was a friendly boy who told good stories and sometimes climbed on a tree stump and pretended to be a preacher or an orator. As he grew tall (by the time he was 17 he was six feet four) he also grew strong, and soon was stronger than anyone about. One day three men were getting poles ready to lift and move a chicken house. Abe picked up the chicken house all by himself. Another time a big bully claimed he could beat up anyone around. He met his match when Lincoln trounced him.

He loved playing jokes. His stepmother Sarah teased him about his height and told him to keep his hair clean so he wouldn't dirty her whitewashed ceiling. So he got some little boys and had them walk with their bare feet in a mudhole. Then he carried them into the cabin, held them upside down, and let them walk on the ceiling. When Sarah saw the footprints she laughed too much to get angry.

Lincoln said she was his best friend in the world. Later, when people told stories about him reading by candlelight, Sarah said it didn't happen often. "He went to bed early, got up early, and then read."

As he grew, he learned to use an ax and to split rails for fences. He was faster than anyone he knew. He worked with his father, cutting timber, sawing lumber, and making things. Sometimes he earned money as a butcher. He built a boat and ferried passengers across a river, until he was arrested for not having a license. That taught him something about the law. For a while he clerked in a store; then he got a job rafting a load of goods down the Ohio and Mississippi rivers to New Orleans. New Orleans was a metropolis of 44,000 people. He was 20, and it was the first time he'd seen a city.

The earliest preserved specimen of Abraham Lincoln's handwriting is in the notebook he kept as a boy and a young man. It says: *Abraham Lincoln, his hand and pen, / He will be good but God knows when.*

6 New Salem

This is the first known photograph of Abraham Lincoln, aged about 40, taken during his term as a U.S. congressman.

Abraham's parents moved again and he helped them move; this time it was to Illinois. But he didn't stay with them. He was 22. His beloved sister, Sarah, had died; he needed to get away, to be on his own. So, with the promise of a job, he walked to the frontier town of New Salem. It had 25 families, and was the same size as another Illinois town: Chicago.

Those who saw him come to New Salem may have looked twice because of his great height and his pants, held up by one suspender and too short by about six inches. If he felt awkward, with his bony arms and legs, and the dark hair that lay like a rough mat on his head, he made up for it by telling jokes and mimicking and clowning around. But he had a serious side; just one look into his deep-sunk dark eyes and you could see that.

He had been trained to do hard physical work—to build, to plow, to plant. He was good at those things, but he didn't like them. It was books and the world of the mind that kept tugging at him. New Salem turned out to be his college: there were opportunities for learning in that little town, and he found them. In New Salem he became a shopkeeper, a postmaster, a scholar, a surveyor, a law student, and a political candidate.

Not all the lessons were easy. Abraham was a partner in a store, and while his partner

He liked to explain to other people what he was getting from books; explaining an idea made it clearer to him. The habit was growing on him of reading out loud; words came more real if picked from the silent page of the book and pronounced on the tongue; new balances and values of words stood out if spoken aloud. —CARL SANDBURG, *ABE LINCOLN GROWS UP*

Joshua Speed, a young Kentucky merchant, moved to Springfield, Illinois, in 1835. He and Lincoln met two years later and shared a room almost until Lincoln married. Josh was Lincoln's closest friend.

The battle of Bad Axe was the last fight in the war against the Sauk and Fox Indians. Lincoln saw no action except with the mosquitoes.

drank most of the profits he sat over a book and ignored the business. When it failed he was left with what he called "the national debt." It took years to pay off; he did it.

He was a friend of the Clary boys, who were rough and wild—hooligans, some called them. But Lincoln would not drink, or cuss, or gamble, even though they did. Perhaps it was his great physical strength that gave him the courage to be different. He was stronger than any of the Clarys, and he proved it more than once.

Soon tales were told of Lincoln's honesty. One day he discovered that he had overcharged a customer by a few pennies; after he closed the store he hiked six miles to return the change.

He walked to the house of the schoolmaster, Mentor Graham (who was well named), and said he wanted to learn grammar. Graham was a born teacher and knew a good student when he found one. They went from grammar to arithmetic to literature to history.

Then the country boy joined the debating society and surprised those who thought he could only joke and tell stories. When he gave a speech it was well crafted and made sense.

After the store failed he had no job, so he enlisted in the army and went off to fight Indians in the Black Hawk War—but he said the only fighting he did was against wild onions and mosquitoes.

During Andrew Jackson's presidency Lincoln was named postmaster of New Salem, even though he supported Henry Clay, who was Jackson's rival. He said New Salem was so small no one bothered about his politics. He put letters in his hat and delivered them. There were no stamps. The person who got the letter paid.

Why was Mentor Graham well named? What is a mentor?

William "Billy" Herndon, who was Lincoln's friend and law partner for 20 years, and his biographer after that. Below is their Springfield law office.

39

In 1847 the Post Office issued the first postage stamps—a five-cent Franklin and a ten-cent Washington.

Springfield may have been the state capital, but its streets were still muddy. On Sundays Lincoln sometimes brought his children to his office (marked with an x), where they would annoy Billy Herndon by pulling books off the shelves and bending the points of his pens.

It was six cents a sheet if delivered within 30 miles. The cost got higher, gradually, for greater distances. Over 400 miles and you paid a quarter. Two sheets cost twice as much, three sheets, three times, and so on.

What the tall young man really wanted to do was study law. In those days there were few law schools. To be a lawyer you studied the law and then passed a test. Abe got some lawbooks and read them, and talked to lawyers and read the books some more. He became a lawyer, a good one. The law would make him wealthy.

His friends urged him to run for public office and he did that, too. The first time he lost, but learned. The next time he got more votes than any other candidate.

In the Illinois General Assembly he spoke only when he had something to say, and that was noted. When the assembly voted 77 to 6 in support of slavery, the young congressman was one of the six. He said, "The institution of slavery is founded on both injustice and bad policy."

Someday he would have much to say on that subject, but first he had more learning to do. He never stopped learning, not as long as he lived.

7 Mr. President Lincoln

An admiring view of Stephen Douglas as a gladiatorial champion of popular sovereignty.

He was smart, no question of that, and honest, and ambitious. Still, he was a rough country boy. How in the world did he get to be president?

Well, it did take some luck, and some hard work, and a lot of public speaking. But mostly it was the nation that was lucky. Because he was the right man for the job—and perhaps people sensed that.

He was against slavery, but he didn't think he could end that terrible practice. He said he just wanted to stop slavery spreading into the western territories. Abraham Lincoln was different from most of those who opposed slavery. He was without malice. He didn't hate the slave owners. Human nature being what it is, he said, southern whites were doing what northern whites would do if they were in their place.

That, however, didn't excuse slavery. Lincoln understood, and said, that "slavery is founded in the selfishness of man's nature—opposition to it, in his love of justice." He knew that those opposing forces, selfishness and love of justice, are found in all people.

Luck came when he ran for the Senate against Stephen A. Douglas. Douglas, who was more than a foot shorter than Lincoln, was called the "little giant." He was a very important man. His suits were made by the best tailors; his friends were influential people. The Illinois Central Railroad lent him a private railroad car so he could campaign in style. They even added a special flatcar with a cannon that boomed to announce his presence. Secretaries and admirers traveled with him as he toured the state of Illinois giving speeches. People paid attention. And

they couldn't ignore his opponent, Abe Lincoln, the candidate of the new Republican Party. Lincoln, dressed in a rumpled suit (showing wrists and ankles), sometimes rode in the same train, but always in an ordinary car and in an ordinary seat. At train stops, Lincoln found Douglas and they debated.

Douglas talked about *popular sovereignty*, the right of the people to govern themselves. (He said that meant the right of the voters to decide if they wanted to have slaves.)

Lincoln said Douglas was hiding from the real issue, which was slavery itself. "The doctrine of self-government is right, absolutely and eternally right," said Lincoln, but that was not the point. "When the white man governs himself, that is self-government; but when he governs himself and also governs another man, that is more than self-government; that is despotism."

Wherever they spoke crowds came; thousands of people poured into small prairie towns. Perhaps they sensed that history was being made.

They were called the Lincoln–Douglas debates—those railroad-stop speeches—and no political contest in American history has ever

Despotism is tyranny, the taking away of freedom, or absolute power over others. A dictator is a despot. So is a slave owner.

Huge numbers of people came to hear the Lincoln–Douglas debates. In Freeport there was a crowd of 15,000. When Lincoln lost the Senate race, he said he felt like "the boy who stubbed his toe; I am too big to cry and too badly hurt to laugh."

THE POLITICAL QUADRILLE
Music by Dred Scott

been as impressive.

Douglas was brilliant; he could have bested almost anyone else, but in Lincoln he met his match. Besides, his cause wouldn't hold up against Lincoln's cold, clear logic. It was 1858, and the South was no longer satisfied with keeping slavery in the southern states. Nor were Southerners really happy with the idea of "popular sovereignty." The Southern extremists—and they were now in control—wanted a slave nation.

They had demanded that Congress pass a fugitive slave law. The law said Northerners had to return runaway slaves to their owners. It made helping an escaped slave a crime. And that made Northerners angry. Now they could not say that slavery involved only Southerners. Now they were involved. Some Northerners were willing to break the law rather than return blacks to slavery. Some Northerners wanted laws passed that would end slavery. They were the *abolitionists*.

Some abolitionists wrote harsh things about the slave owners and the Southern way of life. No one likes to be criticized—especially by people who live far away. So the white Southerners said to the abolitionists, "Mind your own business." Sometimes they said worse things. They hated the abolitionists.

Some white Northerners hated the abolitionists too. They were afraid the abolitionists were stirring up trouble. And, of course, they were. But most Northerners didn't want to face the whole slavery problem; they just wanted it to stay in the South, away from them.

Lincoln made them see the moral issue. He put it in sensible words; he had none of the anger of the abolitionists. He could see both sides. He had been born in a slave state; his wife was a Southerner; but he believed slavery wrong, and said so.

He said black people are "entitled to all the natural rights enumerated in the Declaration of Independence, the right to life, liberty and the pursuit of happiness." And he said:

> *"A house divided against itself can not stand." I believe this government cannot endure permanently half slave and half free. I do not expect the Union to be dissolved—I do not expect the house to fall—but I do expect it will cease to be divided. It will become all one thing, or all the other.*

The 1857 Dred Scott Supreme Court decision—which said essentially that slaves were not people at all, only property—was still a very important election issue three years later, as this cartoon of all the 1860 candidates shows.

Blacks in the North were not treated fairly. Many white Northerners were prejudiced. They hadn't thought much about the words in the Declaration of Independence.

Lincoln's famous phrase, "a house divided," comes from the Gospel of St. Mark in the New Testament.

43

At a Republican campaign rally outside Lincoln's house in Springfield, the tall candidate is visible in a white suit, standing just to the right of his front door. He won—but the bayonets of the seceding southern states were waiting for him.

And:

> *let us have faith that right makes might, and in that faith let us, to the end, dare to do our duty as we understand it.*

Lincoln was heading home on a wet, rainy night when he heard the news that he had lost the election to the Senate. He slipped on the path, got his balance, and said, "It was a slip, not a fall."

Lincoln was no quitter; he would try again.

Stephen Douglas became the senator from Illinois. But the tall, gangling country lawyer was now well known; the Lincoln–Douglas debates had been read across the nation. Two years later, when both men ran for the presidency, people were ready for Lincoln's words. In 1860, Abraham Lincoln was elected president of the United States.

Before he even had a chance to take office, seven southern states seceded from the Union. More would follow.

8 President Jefferson Davis

In his inaugural speech as Confederate president, Jefferson Davis said, "All we ask is to be left alone."

Another Kentucky baby grew up to become a president. He was Jefferson Davis, and he became president of the Confederate States of America. Like Abe Lincoln, Jeff Davis was born in a log cabin in Kentucky.

Jefferson Davis's grandfather (who was four years older than Ben Franklin) came to America seeking opportunity and good fortune. He arrived in Philadelphia, but he didn't find his fortune there, so he moved to Georgia. His son Samuel was born in Georgia.

Sam Davis fought in the Revolutionary War and then, after the war, headed west. He, too, was looking for opportunity. He went west on Daniel Boone's Wilderness Trail, to Kentucky. There he claimed land, took his ax, felled trees, and built a house. He and his wife had 10 children. They went to the Hebrew Bible to name their first sons: Joseph, Samuel, Benjamin, and Isaac. The 10th child was named for the president—Jefferson—with *Finis* as a middle name. That means "the end" in Latin. They meant to have no more babies and they didn't.

The child, Jefferson Finis Davis, was born a year before that other Kentucky baby, Abraham Lincoln.

But soon after Jefferson's birth, Sam Davis and his family moved—to Louisiana—and then again to Mississippi. In Mississippi they found what they were looking for: prosperity. They grew cotton and became rich.

Joseph, the oldest son, became the richest planter in the state. He was an unusual man, Joseph was. He was a lawyer, and kindly, and he owned the finest library in Mississippi. He helped raise his young brother Jefferson and saw that he went to the best schools and to the

Many Christians call the Hebrew Bible the "Old Testament."

When Davis was a lieutenant in the Black Hawk War he mustered in a group of raw recruits. The elected captain of those recruits was a "tall, gawky, slab-sided, homely young man." According to Varina Davis, that man was Abraham Lincoln.

Joseph Davis, Jefferson's older brother, became a second father to him when their own father died. It was Joseph who presented Jefferson with an 800-acre plantation and house called Hurricane (right).

Zachary Taylor, who was Davis's commanding officer—and then, for three months in 1835, his father-in-law.

Military Academy at West Point. He would see that Jefferson had his own plantation and slaves.

Jeff was a mischievous boy, only a fair student, and given to pranks. But he was honest, and honorable, and always true to his word. And he was handsome, with soft blue eyes and a graceful, slim figure.

One day he said he'd had enough of school, he would go no more. His father didn't object, but said he would have to go to work. So Jeff went to the cotton fields; two days later he was back in school.

It wasn't often, however, that he didn't get his way. He grew up, became a soldier, fought Indians in the Black Hawk War, and fell in love with the daughter of "Old Rough and Ready," General Zachary Taylor. Knox Taylor loved him too. Old Zack didn't want the brash soldier boy for a son-in-law. So they eloped—Jeff and Knox—and got married. Then both young lovers caught a malarial fever, and Knox died.

Jake Alker was Jefferson Davis's manservant and stayed loyal to him for the whole of the Civil War.

46

Jefferson and Varina Davis

Jefferson was sick for a long time; when he recovered he was different. He looked different: hollow-eyed and gaunt. He was now serious and studious. He went to his brother's library and became one of the most learned men in the South. He tended his plantation and was a model slave owner, allowing his slaves their own system of justice, with their own court to punish those who did wrong.

And he fell in love again, this time with fair Varina Howell, who made him a good wife. She was 18, he was 36, and the year they married Davis was elected to Congress. Then war was declared, and he went off to fight in Mexico and was a hero. General Zachary Taylor was so impressed with his soldiering he told Davis, "My daughter, sir, was a better judge of men than I was."

Mississippi's governor appointed Davis to the U.S. Senate. Historian Shelby Foote has said that he "was perhaps the best informed, probably the best educated, and certainly the most intellectual man in the Senate." The playful boy had grown into a proper, proud man. But his former father-in-law, Zachary Taylor, changed his mind about him—again.

General Taylor was now President Taylor. He was a southern moderate and a good president. He tried to keep a balance between North and South. It wasn't easy. He said that some southern politicians were "intolerant and revolutionary." Jefferson Davis was one of President Taylor's biggest problems. He was, said Zachary Taylor, the "chief conspirator."

In the Senate, Davis followed the path of South Carolina's Senator John C. Calhoun and argued for Southern power, for the extension of slavery into the Western territories, for the right of a state to secede. Those were the issues that would cause the South to leave the Union. When the Confederate nation looked for a president, it could find no one more qualified than Jefferson Davis.

When Davis's daughter, Winnie, fell in love and wished to marry a fine young man, he forbade it. The young man was a Yankee lawyer and grandson of an abolitionist preacher. Had Davis forgotten his young love? Winnie never married and died 10 years later at age 34.

President Davis as an acrobat, balancing on a rope of unraveling southern cotton and a smoldering time bomb—the Confederate States.

9 Slavery

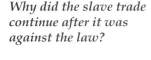

The law that made it a crime to teach slaves was nothing new. Oppressive societies almost always try to censor books and limit learning. In Ireland, British law made it a crime to teach Irish children to read and write the Irish language. Irish children were taught to read and write—but in English.

Slaves about to be sold were fattened up for a few days so they'd look strong and healthy.

Why did the slave trade continue after it was against the law?

Thomas Jefferson, James Madison, and George Washington thought the South would give up slavery on its own. Those Virginia founders were ashamed of slavery. They believed it would gradually disappear.

They were wrong. The slave system became stronger.

But it almost happened. In December of 1831 the Virginia Assembly debated the issue of freedom for the slaves. Virginians were jittery, and perturbed, and agitated. A few months earlier, in August, a slave named Nat Turner had led a revolt. Turner was a preacher who believed God had chosen him to free the slaves. He and his followers killed 55 whites before they were captured and killed. One hundred blacks were slaughtered in the search for Turner. The killings—on

Nat Turner and his followers killed 57 men, women, and children. Troops came in to put down the uprising and killed over 100 blacks, including many who were innocent.

both sides—were too much for decent people to accept.

A Virginia legislator said, "We talk of freedom while slavery exists in the land." Thomas Jefferson's grandson presented Jefferson's plan for gradual emancipation to the Virginia General Assembly. It suggested that slave owners be paid for their slaves.

Many Virginians wanted to free the slaves, but others felt they could not exist without slavery. Many wondered what would happen to the African Americans if they were free. Where would they live? Would they find jobs? Should they be sent back to Africa? The General Assembly's vote was close—73 to 82—but those who were for slavery won. The other side never got another chance.

Southern white slaveholders were now scared of slave uprisings. They knew that in Santo Domingo (Haiti) a brilliant, self-educated slave named Toussaint L'Ouverture had led a successful rebellion that left 60,000 people dead.

Because of the fear, new, harsh laws were passed that made life even more difficult for the slaves. Laws could have been passed to make slavery more humane. Laws could have been passed to prevent the selling of children and the separating of husbands and wives. They were not.

Laws could have been passed to make sexual abuse of slaves a crime. They were not.

Those who beat or murdered blacks could have been brought to trial. They were not.

Slaves could have been allowed to learn to read and write. They were not. Instead, laws were passed making it a crime to teach slaves. Georgia passed a law making it illegal for anyone to write a will freeing his or her slaves. After 1831, slavery became more and more brutal.

Article 1, Section 9, of the Constitution says:

The Migration or Importation of such Persons as any of the States now existing shall think proper to admit, shall not be prohibited by the Congress prior to the Year one thousand eight hundred and eight.

The "Importation of such Persons" means the bringing in of slaves. And "not prohibited" means not stopped. It is a fancy way of saying that the slave trade could not be ended before 1808. As soon as it could, Congress passed a law stopping that awful trade in people. The law went into effect on the first

How many Virginia legislators would have had to change their vote for the General Assembly's result to have come out differently?

Toussaint L'Ouverture ruled the island of Haiti for several years, until 1802, when Napoleon sent troops, abolished the new constitution, and captured Toussaint L'Ouverture.

In a Free State

A Frenchman, Alexis de Tocqueville, visited the United States and observed the difference between a free state and a slave state.

This was what he wrote in 1835: "On the [north] bank of the Ohio, everything is activity, industry; labor is honored; there are no slaves. Pass to the [south] bank and the scene changes so suddenly that you think yourself on the other side of the world; the enterprising spirit is gone."

day of 1808. Thomas Jefferson was then the president. At first, the penalty for slave trading was a stiff fine. Then it became death. But that didn't stop the slavers—there was too much money to be made selling slaves. Soon there was a busy illegal slave trade (just as, in the 20th century, there would be an illegal drug trade). Southerner William Fitzhugh said that slavery was "the natural and normal condition of laboring men," and that the slave trade should be made legal again.

Only about one fourth of the 1.55 million white families in the South actually owned slaves. Of the slave owners, just one in seven had more than 10 slaves. Most Southerners were small farmers without slaves. Some were mountain folk who hardly knew any black people. The big plantation owners—who had hundreds of slaves—were very few in number. But they dominated the South and

One slave, Josiah Henson, describes how he was sold away from his family: "My mother …pushed through the crowd while the bidding for me was going on, to the spot where Riley was standing. She fell at his feet, and clung to his knees, entreating him…to buy her baby as well as herself, and spare to her one, at least, of her little ones….This man disengaged himself from her with violent blows and kicks. …I must have been then between five and six years old."

SLAVERY

ORDAINED OF GOD.

"The powers that be are ordained of God."
ROMANS xiii. 1.

BY

REV. FRED. A. ROSS, D.D.
PASTOR OF THE PRESBYTERIAN CHURCH, HUNTSVILLE, ALABAMA.

PHILADELPHIA:
J. B. LIPPINCOTT & CO.
1857.

An Alabama minister wrote this proslavery booklet in response to abolitionist tracts.

back to the French colonies, former slaves on the island of Haiti led a bloody rebellion and blocked the French effort.

✔ European serfdom (much like slavery) ended in Prussia (part of Germany) in 1806, in the Habsburg empire of Austria–Hungary in 1848, in Russia in 1861, and in Romania in 1864.

✔ America's Civil War lasted from 1861 to 1865 and ended slavery in this nation.

✔ Brazil remained a slaveholding country until 1888, when slavery was ended peacefully.

✔ William McNeil says that "as many as 17 million Africans entered Islamic slavery between A.D. 650 and 1920." For the most part, the slave trade ended in Arab lands when Britain stopped it on the island of Zanzibar (off the east coast of Africa), in 1898.

✗ Until well into the 20th century, some Indians (from India, which was ruled by the British until 1947) worked as indentured laborers in British-controlled possessions in Africa and Asia. They endured conditions as bad as those of many slaves.

✗ Chinese indentured laborers (they had to promise to work for a number of years) worked in the Australian gold rush (1851–1856), helped build the first transcontinental (coast-to-coast) railroads in the U.S. (1860s) and western Canada (1880s), and worked on other projects that needed large numbers of laborers around the world.

Solomon Northup on cotton picking: "Each slave is presented with a sack, which goes over the neck....Each...is also presented with a basket that will hold at least two barrels."

According to a report published in 1993 by the United Nations' International Labor Organization, millions of workers are still enslaved today.

✗ In Thailand, "child catchers" roam the countryside buying or stealing poor children and then selling them in cities as servants or factory workers.

✗ In Haiti, searchers called *buscones* take children and send them to the Dominican Republic to work on sugar plantations.

✗ In Brazil, men called "cats" promise poor people good wages, take them far from their villages, and give them little or no pay. The people have no way to get back home.

controlled the legislatures. Some of them set out to convince people that slavery was a normal and good way of life. "Instead of an evil," said Senator John C. Calhoun, slavery was "a positive good...the most safe and stable basis for free institutions in the world."

Some Southerners (and Northerners, too) believed Africans were "savages" and they were "civilizing" them. They believed the white race was superior. That was a racist idea—although they didn't understand that—and it would lead to evil action wherever it was expressed. It was that same racism that made the settlers treat Indians so cruelly.

Senator Albert Gallatin Brown of Mississippi said slavery was "a blessing for the slave, and a blessing to the master." He said it made

See Book 5 of A History of US *for details of the Compromise of 1850.*

In 1847 the British magazine *Punch* ran this cartoon, *The Land of Liberty*, about the hypocrisy of some Americans' talk of freedom while others were enslaved.

for a much better society than that of the "vulgar, contemptible, counter-jumping" Yankees.

Yankee society was changing its economic base. It was going from an agricultural society to an industrial one. It was the way of the future—but that was hard to see. These were changing times. The South, and its agricultural plantations, had once led the nation. The South had been home to great leaders, and great wealth. Now the North was surging ahead.

Actually, there were two Souths. The Old South of the East Coast states was in trouble. The soil on its plantations was worn out. So were its ideas. Slavery had become inefficient there. The New South, to the west, was filled with cotton plantations that were harvesting riches for a small group of white families—at the expense of black laborers.

The leaders of the Old South blamed Northerners for all their troubles. They couldn't believe that their ideas needed changing. The leaders of the New South were terrified that they might lose their slaves, who were the key to their wealth.

To be a fair society, the South needed industry, railroads, roads, and schools. It needed to give opportunity to all. Some Southerners understood that; most did not. The people of the South followed the wrong-thinking leaders.

But, elsewhere, ideas were changing. Europe's nations were all outlawing slavery. Canada outlawed slavery. In 1826, when the secretary of state, Henry Clay, asked that blacks who had run away to Canada be returned, the Canadian government said, "It is utterly impossible."

Slaves who ran away were called *fugitives*. Now if enslaved African Americans were blessed, as their owners claimed, why were they running away?

By 1850 there were about 30,000 fugitives in

The Liberty of a Man

Words from a letter about the Fugitive Slave Law written on November 21, 1850, to President Millard Fillmore by the abolitionist leader Theodore Parker:

I am not a man who loves violence; I respect the sacredness of human life, but this I say, solemnly, that I will do all in my power to rescue any fugitive slave from the hands of any officer who attempts to return him to bondage. I will do it as readily as I would lift a man out of water, or pluck him from the teeth of a wolf, or snatch him from the hands of a murderer. What is a fine of a thousand dollars, and jailing for six months, to the liberty of a man?

the North—worth some $15 million. Something must be done, said the slave owners. Fifteen million dollars is a lot of money.

A fugitive slave law was passed. It was part of the Compromise of 1850. The Fugitive Slave Law forced Northerners to return slaves to their owners or face criminal charges. It made many Northerners very angry.

In 1854, Anthony Burns, a fugitive slave, was arrested in Boston. He was to be returned to slavery. Bostonians were outraged. On June 2, Burns was led to the city's Long Wharf and put on a boat heading south. A thousand policemen were needed to guard him. Bells tolled, shops in the city were draped in black, mobs howled, "Shame, shame," at the police. That day the government spent almost $100,000—to keep order and return one man to slavery.

This abolitionist poster told the story of Anthony Burns in pictures—his sale, escape, arrest, protest, capture, and imprisonment.

A Massachusetts antislavery poster ran the text of the Fugitive Slave Act and questioned the law's constitutionality.

10 John Brown's Body

In the 1830s and 1840s John Brown befriended fugitive slaves and helped to fund antislavery pamphlets published by blacks.

By 1856 there was civil war in Kansas. Those who believed in slavery were shooting those who did not. It was called "Bloody Kansas," and the nation should have taken warning.

In May of 1856, a tall, fierce-eyed, Bible-quoting white man led a group that brutally murdered five proslavery Kansans. The tall man was named John Brown, and while some people thought him half-mad, others called him a saint. Like Nat Turner (the man who had started a slave rebellion in Virginia), Brown burned with religious fire. He believed he was acting for God.

In March of 1857, the Dred Scott case was decided by the Supreme Court. Dred Scott was a slave; his owner took him into a free state. Scott claimed that made him free. The Supreme Court said it didn't. The court didn't stop with that. It ruled that Scott was not even a legal person; it ruled that he was property and must be returned to slavery.

In Springfield, Illinois, Abraham Lincoln said, "We think the decision is erroneous." Lincoln also said that slavery was "an unqualified evil to the negro, the white man, and the State."

Meanwhile, John Brown had left Kansas and grown a bushy white beard. It was a disguise; he intended to start a revolution. He expected slaves to rise up and join him. One dark, rainy night in 1859, he and 21 followers put rifles over their shoulders, covered the rifles with long gray scarves, and quietly marched into Harper's Ferry in Virginia. Brown had picked that pretty little town as the perfect place to start his slave uprising. It was at a strategic spot: a gap in the mountains where two rivers, the Potomac and the Shenandoah, come together.

Erroneous (uh-RO-nee-us) means mistaken or wrong.

Harper's Ferry was in Virginia in 1859. It was soon to be in a new state: West Virginia.

54

He planned a guerrilla war, and the mountain setting seemed right. Besides, a railroad ran through the town. Even more important, Harper's Ferry was the site of a government arsenal and armory. Guns were made and stored there.

But only one guard stood on duty when Brown and his small army marched into town. The guard thought the men were playing some kind of joke; when he saw rifles pointed in his direction, he knew differently. He was soon a prisoner, and John Brown and his followers had control of the armory. Then a train came by, and that messed up their plans. Disciplined soldiers might have acted calmly, but Brown's men panicked and killed an innocent railroad worker (who happened to be black, and free). Then they let the railroad cars go on their way. That was foolish. The people on the train took news of the uprising to Washington and Baltimore.

Meanwhile, some of Brown's fighting abolitionists were knocking down Colonel Lewis W. Washington's front door. Colonel Washington, who lived near Harper's Ferry, was George Washington's great-great-nephew. He owned a sword that the king of Prussia had given to the first president. John Brown wanted to wear that sword. General Washington had fought for American freedom; John Brown believed

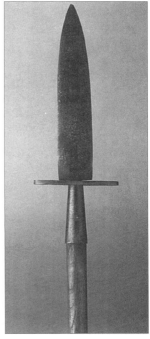

John Brown commissioned a Connecticut blacksmith to make 1,000 of these iron pikes. He planned to arm rebellious slaves with them—but he never got that far. He was stopped at Harper's Ferry (left, showing the railroad bridge across which Brown and his men entered the town).

The night before the raid, John Brown told his men: "We have here only one life to live, and once to die; and if we lose our lives it will perhaps do more for the cause than our lives would be worth in any other way."

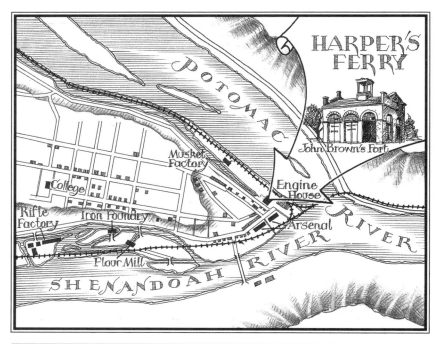

he was doing the same thing. The colonel, the sword, and some townspeople were all hauled off to the armory and held hostage.

Maybe Brown was insane, or maybe he wanted to die for his cause, because, although he had planned the raid for months, he brought no food with him and he kept all his plans a great secret. So the slaves he said would rise up and join him never came. No one knows if they would have come if they'd known what was happening. They didn't learn about John Brown until too late.

It was white Virginians who responded with astonishing speed. Church bells in all the neighboring communities rang the alarm and men on horseback spread the word. Soon, very soon, the town was filled with militia and armed farmers.

Some of them began to drink, and the atmosphere got ugly. Shots rang out. People were killed. One was the kindly mayor of Harper's Ferry. Two of Brown's men were gunned down when they came out

U.S. troops storm the engine house where Brown holed up.

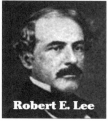

Robert E. Lee

of the armory carrying a white flag.

Then the U.S. Army and the U.S. Marines arrived. They charged into the armory and soon it was all over. Brown was captured—by an army officer named Robert E. Lee—and given a trial that all the nation followed.

John Brown lay in the courtroom on a cot. He said his wounds—sword cuts around the head—made it necessary for him to lie down. Some 500 spectators crammed into the small country courtroom. Peanut shells littered the floors and fingerprints marked the dirty white walls. Outside, on the streets of Charlestown (10 miles from Harper's Ferry), several thousand people stood around hoping to see something of the now-famous man.

Brown pleaded not guilty to treason and murder. His captors didn't realize that the fiery abolitionist had the skills of a brilliant actor. They were giving him an audience that included people all over the country, and in other nations too. Newspaper reporters who covered the trial wrote of his every word and action. John Brown put on a performance few would forget. Often he lied, but he knew how to make people believe in him, and there was truth in the cause he argued. His words inflamed the North:

> I deny everything but…a design on my part to free slaves.…Had I interfered in the manner which I admit…in behalf of the rich, the powerful, the intelligent, the so-called great…every man in this court would have deemed it an act worthy of reward rather than punishment.

The abolitionists were outraged by the trial. They said it wasn't fair. Brown became their martyr hero.

Southerners were outraged too. They were terrified at the idea of a slave uprising. They thought John Brown was a dangerous villain and they wanted him convicted—quickly. They got their wish.

Brown sat on his black walnut coffin in a furniture wagon, and rode to his death. It was a misty Virginia morning, and he looked up at the Blue Ridge Mountains and said, "This is beautiful country. I never had the pleasure of really seeing it before." Then he got down from his coffin—holding his head high, like a proud, beaked, white-tufted eagle—and mounted the steps of the gallows. A square of soldiers surrounded him. One militiaman was a Richmond actor named John Wilkes Booth (remember that name), another was a military-school professor named T. J. Jackson (who would soon be seen standing like a stone wall), and another was Edmund Ruffin (an eccentric old man who later claimed he fired the first shot of the Civil War at Fort Sumter in South Carolina).

Two New England writers said of John Brown:

That new saint will make the gallows glorious, like the Cross.
—RALPH WALDO EMERSON

No man was ever more justly hanged.
—NATHANIEL HAWTHORNE

strangers for a prey, and to the wicked of the earth for a spoil; and they shall pollute it.
22 My face will I turn also from them, and they shall pollute my secret *place:* for the robbers shall enter into it, and defile it.
23 ¶ Make a chain: for the land is full of bloody crimes, and the city is full of violence
24 Wherefore I will bring the worst of the heathen, and they shall possess their houses I will also make the pomp of the strong to cease; and their holy places shall be defiled
25 Destruction cometh; and they shall seek peace, and *there shall be* none.
26 Mischief shall come upon mischief, and rumour shall be upon rumour; then shall they seek a vision of the prophet; but the law shall perish from the priest, and counsel from the ancients.
27 The king shall mourn, and the prince shall be clothed with desolation, and the hands of the people of the land shall be troubled: I will do unto them after their way, and according to their deserts will I judge them; and they shall know that I *am* the LORD.

CHAPTER VIII.

In prison after the raid, John Brown marked passages in this Bible that justified violence to avenge injustice.

**All through the conflict up and down
Marched Uncle Tom and Old John Brown,
One ghost, one form ideal.
And which was false and which was true,
And which was mightier of the two,
The wisest sybil never knew,
For both alike were real.**
—FROM A POEM BY
OLIVER WENDELL HOLMES

Legends, such as that John Brown kissed a slave child on his way to execution, sprang up after his death.

This Question Is Still to Be Settled

While in prison, John Brown told his captors:

I want you to understand, gentlemen…that I respect the rights of the poorest and weakest of colored people, oppressed by the slave system, just as much as I do those of the most wealthy and powerful. That is the idea that has moved me, and that alone.

And then he issued this warning:

I wish to say furthermore, that you had better—all you people at the South—prepare yourselves for a settlement of that question that must come up for settlement sooner than you are prepared for it.…You may dispose of me very easily; I am nearly disposed of now; but this question is still to be settled—this negro question I mean—the end of that is not yet.

John Brown's note to his jailer before execution. It ends, "I had as I now vainly think flattered myself that without very much bloodshed, it might be done."

Brown handed his jailer a slip of paper. It said: *I John Brown am now quite certain that the crimes of this guilty land will never be purged away, but with blood.* He was right.

John Brown stuck his head in a noose made of South Carolina cotton. His ghost would soon haunt both North and South.

Northern soldiers would sing a rousing song about him and Southerners would hate him. Only a few people were able to talk of John Brown with any sense at all.

One of them was Abraham Lincoln. He said, "Old John Brown has been executed for treason against a State. We cannot object, even though he agreed with us in thinking slavery wrong. That cannot excuse violence, bloodshed, and treason."

That winter, Lincoln, who was campaigning for president, said:

> One-sixth of the population of the United States are slaves, looked upon as property, as nothing but property. The cash value of these slaves, at a moderate estimate, is two billion dollars. This amount of property value has a vast influence on the minds of its owners, very naturally. The same amount of property would have an equal influence upon us if owned in the North. Human nature is the same—people at the South are the same as those at the North, barring the difference in circumstances.

And so it came to that. Slaves represented money, and few people will give up their wealth without a fight.

11 Lincoln's Problems

"The hen is the wisest of all the animal creation," said Lincoln, "because she never cackles until after the egg is laid."

South Carolina led the way. It was the first state to secede. Mississippi, one of the richest states in the nation (with more millionaires in the town of Natchez, per capita, than anywhere else in the country), followed eagerly. So did Florida, Alabama, Georgia, Louisiana, and Texas.

The other slave states hesitated until President Lincoln called for volunteers to fight in the South. That decided it for four more states: Virginia, Arkansas, North Carolina, and Tennessee. They would not fight their sister states.

When Virginia joined the rebels the Confederate government was invited to make its headquarters in Richmond. That turned out to be a mistake for Virginia, because much of the war was fought on her land. But no one knew that in 1861. War preparations made Richmond an exciting place: there were jobs to be done and political power to be distributed. Richmond's population increased by one third. The city was an industrial hub. Its iron foundry began pouring out weapons. Workers, soldiers, and white citizens all seemed committed to what they called a "glorious cause." The war preparations had a kind of dreamlike, romantic quality. It was as if the plans were for a game—not the real thing. Few people talked about the side of war that is serious, and deadly. What

Per capita comes from Latin and means "by the head." Considering the number of people in Natchez, the city had a larger percentage of millionaires than anywhere in the country.

South Carolina had a larger proportion of blacks than any other state. More than half (57 percent) of its people were black.

In this 1861 antisecession cartoon a collection of vipers, alligators, wild hogs, and other loathsome beasts is hatching out in the eagle's nest—the Union.

Jefferson Davis decided every Confederate state should be represented in his cabinet (above). So some of the cabinet members, picked more for the state they came from than whether they were any good, were pretty useless.

Broad Street railroad station in downtown Richmond. Above (right) is the White House of the Confederacy, which you can still visit today.

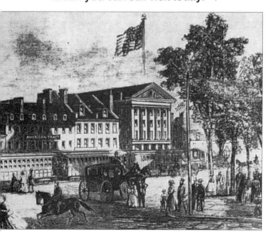

they talked of was heroism and glory. Plays were performed in the city's theaters and fancy-dress balls held in its handsome homes. President Jefferson Davis and his family moved into an elegant, high-ceilinged mansion (to be called the White House of the Confederacy). The columned Virginia capitol (the building Thomas Jefferson designed) was now the Confederate capitol. Most white Virginians were optimistic and determined. Most black Virginians were in no position to act independently.

Actually, not all of Virginia seceded. The people over the mountains, to the west, didn't agree with the plantation owners and the others of eastern Virginia. They were small farmers, and loyal to the Union—so they seceded from Virginia, formed West Virginia, and became the 35th state.

There were four other slave states (besides those that had seceded), and they were undecided. Should they join the Confederate cause? Or stay in the United States? Those slave states were Maryland, Kentucky, Missouri, and Delaware. They were border states—touching North and South. All had influential citizens who wanted to join the Confederacy. Look at a map and see what you think might have happened if they had left the Union. Maryland enclosed Washington, D.C., on three sides. Virginia was on the fourth side.

Could the capital have been saved with Maryland as an enemy? Abraham Lincoln said he needed those border states. He didn't think he could win the war without them.

Control of Kentucky meant control of the important Ohio River. You can move armies and supplies down a river. Abraham Lincoln said he had to have Kentucky.

In addition to vital land, those border states had factories, large populations (to turn into soldiers), and mules and horses. But holding the border states created a problem for Lincoln. If he freed the slaves, the border states might leave the Union.

Meanwhile, the abolitionists were screaming for him to free the slaves. They were calling Lincoln all sorts of nasty names. Lincoln said that if he freed the slaves, lost the war, and destroyed the Union, he wouldn't help the slaves or anybody else.

Besides, he knew that many white Northerners were racists. That was unfortunate (and heartless) but it was 19th-century reality. Lincoln was a practical man and a politician. He couldn't move too fast. He knew the people had to be behind him. There seemed to be two Lincolns. One was the private man who had personal opinions and hated slavery. The other was the president who felt he must reflect the will of the people. And the people had not yet fully decided to abolish slavery. (Harriet Beecher Stowe's book was helping push them in that direction.)

In his inaugural address (the speech he gave when he became president), Lincoln said slavery would be left alone in the slave states. "I have no purpose, directly or indirectly, to interfere with the institution of slavery in the states where it exists," he said. His goal was to preserve the Union. "We are not enemies but friends. We must not be enemies," he said.

President Lincoln understood that slavery was a leftover from another age. He believed that slavery was doomed. Decent people could not tolerate it much longer. To a Philadelphia audience he offered hope that "the weights would be lifted from the shoulders of all men,

Richmond had a large population of free blacks. The war put them in a strange position. Most worked for Confederates, but hoped for the South's defeat. As for the slaves, when they could, many ran away to Yankee army camps. Some remained loyal to the whites they knew. Others waited—to see what would happen.

I think to lose Kentucky is nearly the same as to lose the whole game. Kentucky gone, we cannot hold Missouri, nor, as I think, Maryland. These all against us, and the job on our hands is too large for us.

—ABRAHAM LINCOLN

The border states of Kentucky and Missouri were in an uncertain position. Feeling between those for and against slavery was very strong. Bands of guerrillas, like the Union soldiers (left) firing on Confederate sympathizers in St. Louis, continually came into conflict.

Charleston

Cooper River

Ashley River

CHARLESTON HARBOR

Ft. Sumter

FORT SUMTER

= Confederate Gun Emplacements

Fort Sumter might have held out against a few ships, but it was helpless under fire from the very guns meant to help it protect Charleston harbor.

and that all should have an equal chance." But he would not start a war to end slavery. "The government will not use force unless force is used against it," he said. He knew if there was war, and if the South won, there would be two American nations and continuing battles between them: over runaways and western lands and power. And he knew that the South had a good chance of winning a civil war.

When we look at things today, it seems as if the North should have had an easy time winning. After all, the North had many advantages: more men, more industry, more food. But rebellions are always fought by underdogs. People in power don't try to change things. The outcome of war is often surprising. Our revolution taught us that. And the South was far more powerful than the colonies had been when they fought England.

Like the colonies, the Confederacy wasn't trying to conquer anyone. Its leaders said they just wanted to leave the United States. But that wasn't quite true. They wanted some of those western lands. They wanted to create a slave nation and they wanted it to grow. And they wanted slaves who ran away to free lands returned. But they didn't talk much about that. They just talked about states' rights.

The North had a tough job to do. It had to conquer 11 southern states and then hold them. That would be difficult.

Many Southerners were skilled fighters. They were mostly farmers, used to shooting guns, riding horses, and being out of doors. Many Northern soldiers were city boys who had never shot a gun or sat on a horse. Besides, most of the Rebels believed they were fighting for their land; they were being told that the Yankees were nasty people who wanted to rule over them. Lincoln knew they would not give up easily.

But Lincoln would not let the Confederacy bully him. The Union had forts on the southern coast. One of those, Fort Sumter, was in the harbor at Charleston, South Carolina. As you know, that was where the war began. Here is how it started.

On December 20, 1860, at a state convention, South Carolina voted to secede from—to leave—the Union. It demanded that Fort Sumter be turned over to its new government. U.S. troops were to leave the fort. President Lincoln wouldn't agree to that. Fort Sumter belonged to all the people of the United States, not just to South Carolina, he said. The Carolinians were determined to have the fort. They threatened to starve the men stationed there. Lincoln sent food and supplies. The Rebels fired on the supply ship. Then they fired on the fort—and destroyed it. That was how the war began. It was April 1861.

In the North there were some who urged Lincoln to let the South go. "The Union doesn't need the South," they said. "Who cares about slaves?" they added.

Lincoln didn't agree. But he knew this would be no easy contest. He needed strength and courage if he was to win this war. He also needed good generals. Finding those generals would be his biggest problem.

A Confederate bill or "shinplaster." A shinplaster was a bandage for a sore leg—and that was about all these bills were good for, especially toward the end of the war. By then, even in the South many people would sell goods only for Yankee dollars or gold.

Oil was discovered in Pennsylvania not long before the war started—oil that helped fuel the Northern war effort.

12 The Union Generals

Look at the map on page 67 to understand General Scott's plan.

Do you remember General Scott? He was known as "Old Fuss and Feathers" during the Mexican War.

Cotton Is King

Many Southerners believed that cotton was going to determine the outcome of this war. Without cotton, the English cotton mills would be in big trouble, they said. England had to enter the war on the side of the South, they added. Here are South Carolina senator M. B. Hammond's words (and they were widely believed).

What would happen if no cotton were furnished for three years? England would topple headlong and carry the whole civilized world with her, save the South....No, you dare not make war on cotton. No power on earth dare to make war upon it. Cotton is king!

General Winfield Scott was in charge of the Union army. He was an old man, and in terrible physical shape. He even had to be helped onto his horse. But there was nothing wrong with his mind—it was as sharp as ever.

General Scott looked at the situation when the war began and he figured it would take at least two or three years to win a war against the South. He came up with a plan. To begin, he thought the North should blockade Southern ports. That means Northern ships would patrol the Southern coast and keep ships from entering or leaving. That wouldn't be easy, and a few ships would probably slip through the blockade, but if the South could be kept from trading with Europe it would be in trouble. The Confederacy was an agricultural nation. It didn't manufacture much of anything. If the South was to fight a war, it would need cash to buy goods—especially weapons. To get cash it needed to sell its cotton in Europe. A blockade that closed Southern ports would really hurt the Confederacy.

General Scott also thought the North had to control the Mississippi River. That would cut off Louisiana, Texas, and Arkansas from the rest of the Confederacy. It would close more ports and keep that cotton from being exported.

As to the actual fighting, the Union could send armies from the east and armies from the west to squeeze the Confederacy.

And that is pretty much what happened during the Civil War. But it didn't take two or three years, it took four years: from April of 1861 to April of 1865.

Guess what happened when people heard of Winfield Scott's plan.

Before the battle of Chancellorsville, U.S. General Joseph ("Fighting Joe") Hooker said, "My plans are perfect. May God have mercy on General Lee, for I will have none."

They laughed. The war lasting two or three years? Why, that was the nonsense of an old man. This was a war that would be over in a few months, everyone knew that. A plan to squeeze the enemy? That was just plain silly.

They called Winfield's plan the Anaconda Plan—because an anaconda is a snake that squeezes its prey—and it caused so much dismay that President Lincoln was forced to look for a new general. He found a man who was handsome and intelligent and popular with his troops. He was a graduate of the U.S. Military Academy at West Point and a former railroad superintendent. His name was George Brinton McClellan.

McClellan, who was 35, was a small, dapper man with a strong body, thick dark hair, and a bushy mustache. He rode a black horse named Dan Webster. His admirers called him "the young Napoleon." He liked the comparison with the great French general and often posed, like Napoleon, with one hand in his vest.

McClellan was an excellent organizer. That's an important ability when you are in charge of large numbers of people. A general has to feed, house, and equip his armies. He has to be able to move them long distances. He has to inspire them. He has to train them. McClellan was good at all those things. There was just one problem. It was a big problem for a general. He didn't like to fight. He kept hesitating, and making excuses, and pulling back.

Poor Abraham Lincoln. It is hard enough being president, but being president during a war is really difficult.

> Yesterday the 6th Corps was reviewed by Lieutenant General U. S. Grant. [He] is a short, thick-set man and rode his horse like a bag of meal. I was a little disappointed in the appearance, but I like the look of his eye.
>
> —ELISHA RHODES, UNION SOLDIER FROM PAWTUXET, RHODE ISLAND

General McClellan (above) said to the men of the Army of the Potomac: "I am to watch over you as a parent over his children; and you know that your general loves you from the depths of his heart."

Outnumbered?

McClellan's Union forces outnumbered the Confederate Army three to one. Yet he was convinced he was outnumbered and refused to fight. Edwin M. Stanton, Lincoln's secretary of war, said of him, "If he had a million men, he would swear the enemy had two millions, and then he would sit down in the mud and yell for three."

Once Lincoln ordered McClellan to advance immediately. "It was nineteen days before he put a man over the river [and] nine days longer before he got his army across, and then he stopped again," said Lincoln, who fired the general. "They have made a great mistake," said McClellan, on hearing the news. "Alas, for my poor country."

Lincoln's general U. S. Grant (above, at the battle of Cold Harbor) once said, "I don't know whether I am like other men or not, but when I have nothing to do, I get blue and depressed."

Especially when you can't find a good general. Lincoln tried Generals McClellan, Frémont, Burnside, Halleck, Hooker, Pope, Meade, and some others. Not one of them fought the way Lincoln thought they should. Every time the North lost a battle—and it lost quite a few—Lincoln got blamed; even though the generals weren't doing what the president was asking.

There just didn't seem to be a leader he could trust. Then he looked out west and there was a general who was winning battles. A general who trapped a whole Confederate army and took all the soldiers prisoners of war. That general, who was good at fighting, was named Ulysses S. Grant. (He even had initials to match his country.) Lincoln called him "the quietest little fellow you ever saw."

Grant had been to West Point, where he was nicknamed "Uncle Sam" because of those initials. His army friends called him Sam Grant. At West Point Sam Grant was a fair student but too small to excel at any sport but riding. He fought in the Mexican War, then left the army, and wasn't much of a success in civilian life. He was poor, really poor, when he inherited a slave. He could have sold the slave for $1,000, but he didn't do it. He gave the man his freedom.

It didn't look as if Grant would amount to anything, until the Civil War came along, and it made him famous (famous enough to one day be president). He was the kind of general who didn't worry much about military theories. He just outkilled or outlasted his enemy. General Grant and his friend, red-bearded General William Tecumseh Sherman (whom we will get to later), were just the kind of generals Lincoln had been looking for.

The owners of blockade runners (fast steamers that slipped past Union warships to bring food and other goods to the blocked-off South) were usually in it for the money (not out of patriotism). At right, a midshipman on a U.S. frigate.

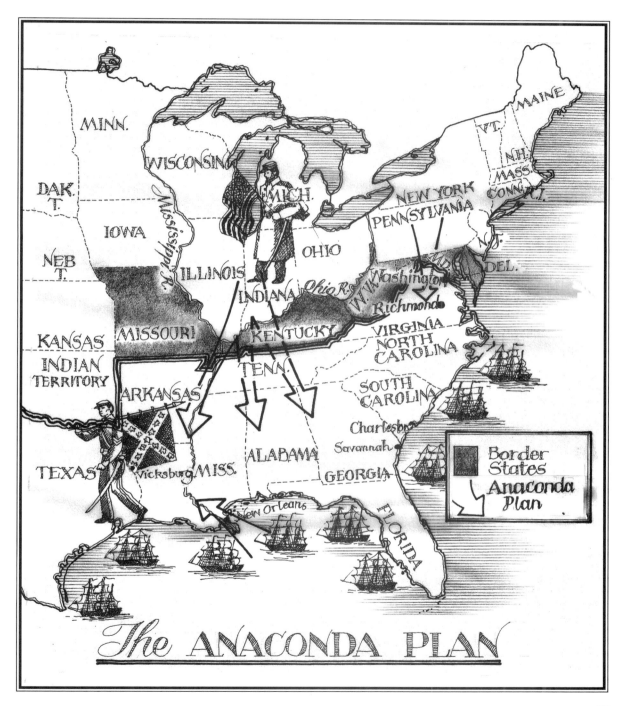

The ANACONDA PLAN

13 The Confederate Generals

A Confederate soldier and his young brother. "I'd rather die than become a slave to the North," said one soldier.

Southern boys liked to play soldier. When they grew up many of them went to military academies. Because of that soldiering tradition, the South had good generals—lots of good generals. To begin, there was James Longstreet. "Old Pete" was what his men called him. He was a big man, with bones that stood out like those of an old hunting dog. Longstreet was cold-eyed, and looked like what he was—a fighter. He usually wore old clothes, dirty boots, and a black hat. While a battle raged, you might see him whittling on a stick. He knew how to act calm. But no one ever questioned his bravery. When he was needed, he was in the front line, leading charges.

Then there was J. E. B. "Jeb" Stuart, who wore elbow-length white gloves, a red-lined cape, gold spurs, and an ostrich feather in his hat. He wanted to look like an English cavalier, and he did. Stuart was one of the most daring cavalry leaders the world has known.

And there was George Pickett, another man who dressed like a dandy, wore his hair long and curled, and put a feather in his broad-brimmed hat. Pickett was at the bottom of his class in the Military Academy at West Point. It didn't matter; he did well enough on the battlefield. Besides, everyone liked George Pickett. You couldn't help

After the war was over, George Pickett went home to Richmond and sold life insurance until his death in 1875.

liking him—he was easygoing, good-natured, and friendly. He didn't drink or gamble, because he had promised his girlfriend, Sallie, that he wouldn't, and he stuck to his word. More than anything, Pickett wanted a chance to fight and show how brave he was. He would have that chance.

The greatest of the Southern generals were Thomas J. "Stonewall" Jackson and Robert E. Lee.

Jackson was an orphan child who grew into a strange, brooding man. Like Grant and Lee, and most of the other generals on both sides of this conflict, Jackson went to West Point and fought in the Mexican War. He was a teacher at the Virginia Military Institute when this war began. He wasn't popular. He was strict and had no sense of humor. The boys called him "Tom Fool" behind his back.

It was different when he took command. His troops were in awe of him. Although, sometimes, it was hard to understand why. Jackson kept apart from his fellow officers and was stern with his soldiers. But he was a winner, and soon everyone knew it. He was intelligent, and daring, and—perhaps because he had a strong religious faith—nothing scared him.

Stonewall Jackson kept doing things that couldn't be done. He marched his men farther and faster than armies could march, beat forces much larger than his, and won battles that were said to be unwinnable.

In 1862 Major General J. E. B. Stuart took over command of the entire Confederate cavalry. General Robert E. Lee called Stuart "the eyes and ears of my army."

The headline in a Richmond newspaper said: "Glorious old Stonewall is fast becoming the HERO OF THE WAR."

In the North, mothers sometimes threatened their naughty children, "Be good or Stonewall will get you!"

"I can anticipate no greater calamity for the country than a dissolution of the Union," said Robert E. Lee (above, as a lieutenant, not long out of West Point), "and I am willing to sacrifice everything but honor for its preservation."

"Always mystify, mislead, and surprise the enemy," said Stonewall Jackson. "And when you strike and overcome him, never let up in the pursuit." This picture was taken around the time of the Mexican War.

Once he captured 400 Federal railroad cars. He threw most of them into a river, but some he had hitched to horses and pulled to Southern railroad tracks.

If you'd seen Jackson, you wouldn't have been impressed. He was awkward and rumpled, and he liked to sit on his dumpy little horse, Sorrel, sucking a lemon. His face was angular and bearded, like a patriarch from the Bible. His men compared him to Joshua, the great biblical commander. They followed him, did what he asked, and knew that he would accept nothing but victory.

Stonewall was an unflinching fighter who saw the hand of God in everything he did—and much of what he did seemed miraculous to most mortals. In Virginia's Shenandoah Valley, he marched a small army over 400 miles (in just over a month), kept a large Federal army off balance, seized much-needed supplies, inflicted heavy casualties, and inspired legends and a feeling of invincibility in his troops. "Boys," said one Union general, "he's not much for looks, but if we had him, we wouldn't be in this trap."

Robert E. Lee was different. He did look as a general should look: handsome, gray-haired, and dignified. He sat erect and unruffled on his beautiful horse, Traveller, and commanded as a general should command: with fairness, audacity, and courage. Men were awed by him, and rushed into battle and died for him. If ever there was a born leader, it was General Robert E. Lee.

His wife was a granddaughter of Martha Washington. His father was "Light Horse Harry" Lee, a Virginia planter and Revolutionary War cavalry hero. Light Horse Harry was said to be fearless in battle, and Patrick Henry didn't scare him either. When Patrick Henry was trying to get Virginians to vote against ratification of the new Constitution, Harry Lee spoke up. This is part of what he said:

> The people of America, sir, are one people. I love the people of the North, not because they have adopted the Constitution, but because I fought with them as my countrymen....Does it follow from hence that I have forgotten...my native state? In all local matters I shall be a Virginian: in those of a general nature, I shall not forget that I am an American.

Those words must have worried Robert E. Lee, because Lee loved and admired his father (even though he had died when Robert was still a boy). Robert E. Lee, too, loved America and he loved people of the North as well as those of the South.

He didn't like slavery, and he freed his slaves before the war ended.

I cannot raise my hand against my birthplace, my home, my children.
—ROBERT E. LEE

And he didn't think much of states' rights, either. So it wasn't easy for him to join the Confederate cause. In a letter to his son, Lee wrote:

> Secession is nothing but revolution. The framers of our Constitution never exhausted so much labour, wisdom and forbearance in its formation, and surrounded it with so many guards and securities, if it was intended to be broken by any member of the Confederacy at will....Still, a Union that can only be maintained by swords and bayonets, and in which strife and civil war are to take the place of brotherly love and kindness, has no charm for me. If the Union is dissolved, the government disrupted, I shall return to my native state and share the miseries of my people. Save in her defense, I will draw my sword no more.

The Great Stonewall

*H*e would never mail a letter that would be in transit on Sunday. He was a strict observer of the Sabbath. And yet so many of his battles were fought on Sundays that the soldiers began to believe that he would fight on Sunday because the Lord would be even more with him....They said he was the most ridiculous figure on horseback. He looked like somebody of oddly assorted parts put together and all jiggling in different directions when he rode.

Jackson is an eerie character; an Old Testament warrior who believed in smiting them hip and thigh. After the bloody fighting at Sharpsburg...Jackson was sitting on his horse eating a peach...and he looked out over this field where there were dead of both sides littered all over the place. And as he was eating the peach he said, "God has been very kind to us this day."

—SHELBY FOOTE

*T*here is a magnetism in Jackson, but it is not personal. All admire his genius and great deeds; no one could love the man for himself. He seems to be cut off from fellow men and to commune with his own spirit only, or with spirits of which we know not.

—CHARLES BLACKFORD, CAPTAIN, CONFEDERATE CAVALRY

He would have a man shot at the drop of a hat—and he'd drop it himself.

—SAM WATKINS, CONFEDERATE SOLDIER

His name was a terror in the Union army, and with us expressed more fear than all other names put together.

—AUSTIN STEARNS, PRIVATE, 13TH MASSACHUSETTS

Stonewall Jackson rallying his soldiers at Chancellorsville in Virginia. An unusual aspect of the battle was that the armies fought into the night, something they generally avoided.

71

I have fought against the people of the North because I believed they were seeking to wrest from the South its dearest rights. But I have never cherished toward them bitter or vindictive feelings, and I have never seen the day when I did not pray for them. —ROBERT E. LEE

Robert E. Lee on his famous horse Traveller. "So great is my confidence in him," said Stonewall Jackson of Lee, "that I am willing to follow him blindfolded."

When the war began, Lee was thought to be the outstanding officer in the U. S. Army. President Lincoln offered him command of the whole Union army. It was hard to turn that job down. His wife said he stayed up all night walking back and forth trying to decide what to do. He had been a loyal officer in the United States Army. Still, when he had to make a choice, he chose Virginia. It is difficult for us to understand today, but many, like Lee, were above all devoted to their home states.

Some say Lee was the finest general America has produced. Perhaps—although the men who fought for him died in terrible numbers. He took bold chances and when they worked, as they often did, he seemed touched by genius.

Few generals have ever inspired people as General Lee did. Few people convey the integrity and intelligence and decency that Lee did. On the battlefield he was cool and daring, but it was in defeat that he showed the best of himself. When the war was over he refused to be bitter, or angry, or anything but noble. He was a symbol to the South of all that was good in themselves and to all Americans he became a heroic figure.

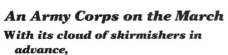

An Army Corps on the March

With its cloud of skirmishers in advance,
With now the sound of a single shot snapping like a whip, and now an irregular volley,
the swarming ranks press on and on, the dense brigades press on,
Glittering dimly, toiling under the sun—the dust-cover'd men,
In columns rise and fall to the undulations of the ground,
With artillery interspers'd—the wheels rumble, the horses sweat,
As the army corps advances.
—WALT WHITMAN

14 President Davis's Problems

President Davis understood military matters better than economics and finance, and his government suffered as a result.

President Davis had superb generals; everyone agreed on that. What he didn't have enough of was food, clothing, weapons, and ships. Slavery had kept the South feudal and agricultural while much of the world was turning industrial. Poor planning had made the region dependent on just a few crops. Southern land was used mostly for growing cotton, tobacco, rice, or sugar. During the war, meat, vegetables, and grains were in short supply. The Confederacy's railroads weren't as good as those in the North. Poor transportation made it difficult to get food to where it was needed. Poor management made things worse than they had to be.

It isn't easy to fight when your stomach is empty, but Rebel soldiers—fighting in an agricultural region—were often half-starved. It isn't easy to fight when you have no shoes, but Rebel soldiers were

I want to live in no country in which the man who blacks my boots and curries my horse is my equal.
—LOUIS T. WIGFALL, SENATOR FROM TEXAS

In a *feudal* society, the laborers are required to work for a master. In return the master is supposed to protect them from enemies and give them enough land to support themselves.

Shoeless Union bodies after battle. A dead man's boots were fair game.

73

The Articles of Confederation were our first constitution. Written in 1777 (during the Revolutionary War), the Articles created a confederation (1781–1788) of 13 states that were almost like 13 separate nations. Some people, who talked of "states' rights," preferred that kind of constitution, but most Americans didn't. In 1788 the Articles were replaced with a whole new constitution creating a federal republic. That meant that the central government was a strong one.

The fat man is Britain, trying to decide whether to back the Confederacy and so keep getting cotton for its textile mills, or to behave decently and oppose slavery.

sometimes forced to take shoes from the feet of dead Union soldiers.

President Davis and the Confederate leaders weren't worried. They were absolutely sure: England would provide supplies. After all, England had always bought southern cotton and tobacco, and Southerners had always bought manufactured goods from England. English aristocrats and southern plantation owners were good friends. Many Englishmen and women didn't like Yankee ideas on equality. Besides, England needed cotton for her cotton mills. There was no question about it: England would come to their rescue.

The Southern leaders had made a big miscalculation. It just happened that English warehouses were filled with cotton. There was a surplus on hand. And, when the Union navy blockaded Southern ports, the British navy didn't like the idea of interfering with a blockade.

The world was changing, and England too. Slavery was now recognized as evil. Great Britain had abolished slavery in 1772, but it was Harriet Beecher Stowe's book that

Charleston's docks were piled high with cotton bales before and after the war, but not while it was going on. The strategy was to export nothing until other countries recognized the Confederacy.

made the British understand that slaves were real people. A million and a half copies of *Uncle Tom's Cabin* were sold in England in one year. It was astonishing: everywhere people were weeping over Uncle Tom and Eliza. Even the queen wept, or so it was said. Some British politicians wanted to help the South, but most of England's people wouldn't hear of it; they had become ardent abolitionists.

President Davis had another problem. Remember when our country was governed under the Articles of Confederation? Remember that a confederation hadn't worked well because each state had more power than the central government?

Well, the Confederacy was a confederation. President Davis just didn't have the political power he needed to fight a war. He couldn't make the Confederate states do anything they didn't want to do. Like pay taxes.

He had yet another problem. It was his personality. He was honest and could always be trusted, but he was also stubborn and irritable. People didn't enjoy working with him.

And he didn't write well. Yes, he knew where to put his commas and semicolons; he'd had a fine education. But he couldn't express himself as Abraham Lincoln could. He wasn't eloquent. He wasn't inspiring. That didn't help his cause.

Mary Chesnut

Republics—everybody jawing, everybody putting their mouths in....republics can't carry on war.... Hurrah for a strong one-man government.
—Mary Chesnut, South Carolina

Davis tried to get help (here he is offering southern cotton and bonds) from France's emperor Napoleon III (nephew of the first Napoleon). If Britain or Russia had backed the Confederacy Napoleon probably would have too—but he didn't want to look like the only European country supporting rebels fighting to keep slaves.

15 Choosing Sides

Virginian William Presgraves didn't even have a proper uniform. He died in 1862, a year after this picture was taken.

It was more than a war that split the nation. It was a war that split families, too.

Yes, brothers really fought brothers. Major Clifton Prentiss—Union army—and his younger brother, William—Confederate army—both fought and died in the same battle at Petersburg, Virginia.

Four of Abraham Lincoln's brothers-in-law fought for the Confederacy; three died for it. Three grandsons of Henry Clay fought for the Union, four for the Confederacy. In one battle, Confederate cavalry general J. E. B. Stuart was chased by Union cavalry general Phillip St. George Cooke, who was Stuart's father-in-law. (General Stuart said General Cooke would regret being a Union man "but once, and that will be continuously.") Senator John Crittenden's two sons became generals: one for the South, one for the North.

Most men went to war for their region. But some believed in the cause of the other part of the nation and fought for those beliefs.

Admiral David G. Farragut was a naval commander (a great one) on the side of the

A young Union volunteer from Michigan. His uniform consisted of a side knife and a revolver.

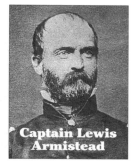

Captain Lewis Armistead

North, though he was from Alabama. Robert E. Lee's cousin, Samuel P. Lee, commanded the Union naval forces on the James River. Generals Winfield Scott and George H. Thomas, both Virginians, fought for the Union.

Winfield Scott Hancock (seated) became a general before the war was over and was very popular with his troops.

Choosing sides was not easy. Two army officers were stationed at a small western outpost named Los Angeles when war was declared. They were outstanding men. Both were natural leaders: strong, kind, and courageous. They were best friends. One, Captain Winfield Scott Hancock, stayed in the Union army. The other, Captain Lewis Armistead, became a Confederate. When they parted, Armistead looked at Hancock and said, "You'll never know what this is costing me, but goodbye, goodbye."

They were to meet again, two years later, at a ferocious battle near a little Pennsylvania town named Gettysburg. What happened? Well, keep reading and you'll find out.

Did they hate each other, Northerners and Southerners? Often they thought they did, but when they got to know each other it wasn't so. There were too many stories of friendliness for the hatred to be real. It wasn't unusual for Northern soldiers to camp in trenches facing Southern soldiers while both sides waited for orders to fight—and that could take weeks or months. At first they yelled at each other, then they talked, occasionally they even sang together.

Once some Yankee soldiers were invited to a dance by some Rebels. The Yankees went and had a good time until an officer found out, came for them, and made sure they never did that again.

Sometimes they traded things, or exchanged letters. Then, when the orders came, they would settle down to killing each other. After all, that's what war is about.

But they did have beliefs, and most of them knew in their hearts that they were fighting for things they believed in. Yes, they knew it had something to do with slavery, and where they lived, and states' rights *vs.* national unity, but there was something more than that. They were fighting, on each side, for a way of life. The southern way of life was different from the northern way.

The people of the North had set out to do something no people had ever done before. They weren't fighting for land. They weren't fighting for riches. They were fighting to make other men, women, and children free.

Many West Point cadets (above) found themselves on opposing sides. Less spiffily dressed, but just as confident, were the men of the Washington Artillery of New Orleans (above right), who had fought bravely in the Mexican War.

The fact of slavery upon this incredibly beautiful new clean earth was appalling, but more than that was the horror of old Europe, the curse of nobility, which the South was transplanting to new soil. They were forming a new aristocracy, a new breed of glittering men.

—MICHAEL SHAARA,
THE KILLER ANGELS

The North was becoming urban. There were cities and factories and all the problems and opportunities that come with industry and change.

In the South, life was pastoral, conservative, and orderly. It was where the old European class society had taken root in the New World. The South offered great opportunity and ease for a small number of white people. For many other white people living was comfortable and secure. But for poor whites and blacks it wasn't comfortable at all.

Life in the antebellum South was not fair. Everyone is not created equal, or given an equal chance, in a slave society.

In a slave society some men and women are treated like prisoners, even though they have committed no crime. In some southern states, like South Carolina, more than half the population was black. Blacks were the prisoners. They did the hard work in the South so that life could be easy for others.

The North was fighting for democracy. Democracy is rarely orderly, but it

Edwin Francis Jennison was a private in a Georgia regiment. He was killed at Malvern Hill, Virginia, in 1862. He was then aged 16.

A lot of Rebels, like this young man, took their slaves to war with them as servants. In most states, it was the first time that southern blacks had ever been allowed to carry arms.

attempts to be fair. Northern factory workers and farm laborers were not rich; sometimes their lives were very difficult, but they were free and better off than workers almost anywhere else in the world.

Virginia's Senator James M. Mason understood the differences between North and South. He said, "I look upon it [the Civil War] then, sir, as a war of... one form of society against another form of society."

So if someone asks you if the war made a difference, you can say yes. It not only ended slavery and preserved the Union, it settled that democratic question. After the Civil War, the United States was committed by constitutional amendments to democracy—to fairness—to equality of opportunity.

Did that mean that everyone was now treated fairly? No, it didn't. It is easier to change laws than it is to change ideas and habits. The fight for real equality of opportunity continued. It continues today. But the goal of fairness was established in the law. The flaw in the Constitution that allowed slavery was amended. No one could ever again argue that slavery was a positive good. Now it was clear: tyranny and persecution and bigotry were forbidden by the Constitution. They were un-American.

Union and Confederacy go at it like gladiators as Caesar, a slave, presides from a throne made of cotton bales. Which gladiator do you think the cartoonist sympathized with?

16 The Soldiers

**Take your gun and go, John,
Take your gun and go.
For Ruth can drive the oxen,
And I can use the hoe.**
—CIVIL WAR SONG

For many proud Southerners, conscription (being drafted or made to serve in the army) was an indignity. Soldier Sam Watkins wrote, "From this time on…a soldier was simply a machine, a conscript.…All our pride and valor was gone.…[The law] allowing every person who owned twenty negroes to go home …gave us the blues. We wanted twenty negroes."

Conscription was looked on in the South where they had it first as a complete downfall of democracy, freedom, and everything else.…But of course it was impossible to fight without conscription.
—SHELBY FOOTE

Drummers like John Clem didn't just beat calls to muster and meals. In battle the drums told the troops what they should be doing.

Their median age was 24. That means that half of the Civil War soldiers were younger than 24, and half were older. Many were 18 or 19. Some were even younger. Eleven-year-old Johnny Clem was a drummer boy in a Michigan regiment. When a Confederate colonel tried to take him prisoner, Johnny picked up a musket and killed the colonel. He was made a sergeant.

The Union rules said a soldier was supposed to be at least 18 years old. But boys who were eager to fight found a way around that. They wrote the number 18 on a piece of paper and put it in their shoe. Then, when asked, they could say, "I'm over 18."

Young men weren't the only ones anxious to fight for their beliefs. Iowa had a famous "Graybeard Regiment." Everyone in it was over 45. "So old were these men, and so young their state," wrote historian Bruce Catton, "that not a man in the regiment could claim Iowa as his birthplace. There had been no Iowa when these Iowans were born."

At first there were so many volunteers that neither army could handle them all. Later, when the volunteers wrote home about the battles and the deaths and the conditions, fewer came willingly, so the governments paid cash rewards for volunteers and finally both sides had to draft men. (Which means they had to force them to sign up.)

Some people called it "a rich man's war and a poor man's fight." That was because, if you were rich enough, you didn't have to be in the army. Confederates who owned 20 or more slaves could get excused, although many fought anyway. And Northerners who could afford the cost were allowed to pay someone to fight for them. Many did.

Most soldiers were farmers, because it was a country of farmers. Some were small-town boys; most had never been far from home. Soldiering sounded like an adventurous thing to do, and for a while it was a bit like boys' camp. One Illinois recruit wrote in a letter home, "It is fun to lie around, face unwashed, hair uncombed, shirt unbuttoned and everything un-everythinged."

A boy's life in those days was different from yours. There were no national sports like baseball or football. Some played a ball game known as "one old cat" with three bases. It was a popular form of bat and ball that changed a bit until it became baseball. The soldiers had lots of time to play ball games in their army camps. After the war, they took the new baseball rules home with them and taught them to their friends and children. Besides ball games, most children played croquet, tag, marbles, hopscotch, and a game called shinny, which involved hitting a tin can or a block of wood toward a goal. Most grown people worked during the daylight hours; they didn't have much time for recreation or travel. So joining the army and getting away from home seemed an exciting thing to do.

For the soldier boys there were new friends and uniforms and parades and drills—but that soon changed. Then, often, there were long

These young men posed for their pictures with slabs of hardtack in hand. "Soft bread" was a delicacy not often met with in the ranks.

A Union encampment. The 96th Pennsylvania regiment is drilling in columns of companies, which are smaller sections of a regiment. The regimental commander was a colonel; companies were led by captains.

A ***breech-loading*** gun is loaded at the bottom of the barrel, near the trigger, as most rifles are today.

marches, long, boring encampments, homesickness, bad food, hunger, and disease. For every man who died in battle, two died of sickness. (Few people then understood about germs and the importance of good food and cleanliness, or how to deal with epidemics.)

Many of the new soldiers—especially town-bred Northerners—didn't know how to handle the rifles they were given. Some were thrown into battle without training—they had to learn under fire. Many of their officers were volunteers, too, and knew no more than the men they led. If the officers weren't natural leaders they were in trouble. As one Indiana soldier wrote, "We had enlisted to put down the rebellion, and had no patience with the red-tape tomfoolery of the regular service....a soldier was ready at the drop of a hat to thrash his commander—a thing that occurred more than once."

They were armies of independent men—on both sides —and the leaders who handled them best understood that. But they could fight like fury, and they fought so well and died so bravely that it is hard not to weep at the thought of their valor.

Northern armies were well fed and well supplied; Southerners often went hungry. Both sides usually had guns enough, although they were not always the best available. An inventive Yankee named Christian Sharps spent six years developing a breech-loading rifle. It was a single-shot weapon, but much better than most guns in use. Abraham Lincoln examined a Sharps rifle, loaded and shot it, and thought it a "wonderful gun." Some new weapons could be fired more than once without reloading. The Spencer seven-shot was the best of the repeating rifles.

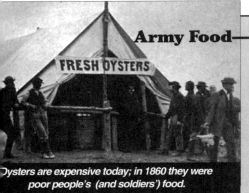

Oysters are expensive today; in 1860 they were poor people's (and soldiers') food.

Army Food—Yum? No—Yuck!

In the Southern army you ate something called "sloosh." You got issued cornmeal and bacon and you fried the bacon, which left a great deal of grease in the pan. Then you took the cornmeal and swirled it around in the grease to make the dough and make a snake of it and put it around your ramrod and cook it over the campfire. That was called sloosh. They ate a lot of that. —SHELBY FOOTE

In the Union army, there was usually food enough—at least between battles. "We have abundant rations, plenty of good water, the river to bathe and wash clothes in....Soft bread is fur-nished daily....Blackberries are plenty. Having enough to eat and little to do, we enjoy ourselves."

When they were on the march, the food wasn't always as abundant. "Some of those men who had been so weak that they could hardly wag a tail, or keep anywhere near the column all day, recovered their strength and spirits and went out foraging after the rest of us had lain down to rest. The squalling of hens...announced their arrival at a farmhouse and that 'fowl' play was going on." —PRIVATE JOHN W HALEY, 17TH MAINE

Tell Ma I think of her beans and collards often and wish for some. But wishing does no good. —BENJAMIN FRANKLIN JACKSON

It was no uncommon occurrence for a man to find the surface of his pot of coffee swimming with weevils after breaking up hardtack in it...but they were easily skimmed off and left no distinctive flavor behind. —BILLINGS, UNION SOLDIER

Richard Gatling devised a machine gun that fired 250 rounds a minute from six revolving barrels. Before the end of the war, several thousand Union soldiers were using new weapons.

But most soldiers used rifled muskets that were loaded from the shooting end (the muzzle). A rifled musket has a barrel with grooves inside that make the bullet spin and go farther and with more accuracy than bullets shot from a plain barrel. Very skilled soldiers could get off three shots a minute.

The new rifled guns were much better than the muskets used in the Revolutionary War or in the Mexican War. General Ulysses. S. Grant, who had fought in Mexico in 1846, said of the old short-range muskets, "A man might fire at you all day without your finding it out." During the Revolution, most of the fighting had been eyeball to eyeball, and many deaths were from bayonet wounds. That wasn't true in the Civil War. Rifled guns could kill a man a half mile away. There were new bullets, too, and they were killers, and cannons—called artillery—that were much more powerful and accurate than those used in Mexico. Much of the fighting, however, was still done in the old way. Most of the generals had gone to West Point, and had learned battle strategy by studying the old battles. The old lessons didn't work with new weapons, and that was part of the reason for the incredibly high numbers of battlefield deaths.

Picture two sides fighting: on one side are the attackers, on the other the defenders. The defenders line their men up in two rows. The front row is kneeling, guns in position. Behind them men stand ready to shoot over their heads.

Here come the attackers. Shoulder to shoulder they charge in great waves of men. They march

Above, from bottom to top, increasingly efficient firearms: a converted flintlock musket, a breech-loading Sharps carbine, and a Spencer repeating rifle (below left are the Spencer's cartridges and box). These were so much more accurate than earlier weapons that hand-to-hand bayonet fighting of the kind shown above (at Williamsburg in 1862) was quite rare. Soldiers were more likely to be killed by sharpshooters' bullets or cannonballs. And, even more likely, by disease.

83

Union general Daniel Butterfield didn't like the bugle call that was played to tell soldiers to turn their lights out. So he scribbled some musical notes on the back of an envelope and gave them to his bugler. It was 1862, and the tune was called "Taps." It was soon being played by both armies.

The machinery of war—artillery and cannonballs—waiting for action. The Union armies often had nearly as many men guarding supplies as fighting at the front; even so they lost huge quantities of arms to Confederate raiders.

to the beat of a drummer boy. The beat tells the men when to stop to fire and reload.

To that picture of attackers and defenders you need to add some soldiers on horseback with swords flashing. And something else: noise. The explosions of cannons and muskets, and the din of men yelling and horses snorting. You can add something else again. Smoke. Lots of smoke. Smokeless powder hadn't been developed, so the battlefield was soon so thick with smoke—especially from cannons—that no one could see what was going on. It isn't surprising that many men were killed by shots from their own side.

How does the battle turn out? Usually the attackers get mowed down. (When the defenders dig trenches and protect themselves—which happened for the first time in the Civil War—the defenders are clearly in the best position.)

It is easy for us, today, to understand that. It wasn't so easy for the generals back in the 19th century. In the old days of muskets, before the Civil War, when bullets weren't as deadly, or guns as accurate, the attackers usually had the advantage. They could charge across an open battlefield and overwhelm the defenders with bayonets and swords. It took time for the Civil War generals to adjust to the new weapons and technology.

Besides, armies of volunteer civilians fighting for a cause just don't fight the same way as professional soldiers. They fight harder. The generals had to learn that too.

And they had to learn the importance of new transportation and communication methods. With railroads, armies could be moved with astonishing speed. With the telegraph, you could instantly find out what was going on hundreds of miles away.

During the Civil War, for the first time in America, war took to the air: hot-air balloons went aloft carrying spies in floating baskets who peered down over enemy lines. The balloons were tethered—that means they were tied to the ground with long ropes. One balloon broke its tether and floated away with a general aboard; luckily, for him, the wind shifted and brought him back to his own lines. Balloons weren't the only Civil War novelty: a submarine sank an enemy ship—and got sunk itself in the process.

When war began, battles were still expected to be controlled fights between armies. Remember how

Balloons were first used in wartime in France during the French Revolution. In the Civil War, Thaddeus Lowe went up in this hydrogen balloon to reconnoiter for the Federal army. But the Confederates were unable to find the resources to field their own balloons.

Earthworks—above, a powder store—and engineering skills became very important in the Civil War. Union engineers built this bridge over the Potomac in under 40 hours.

civilians brought picnic baskets to see the battle at Bull Run? They expected a by-the-rules, orderly skirmish. But in this war all the rules changed. It has been called the first modern war. Cities and farms were burned and civilian populations terrorized. It was total war, and it got out of control.

17 Willie and Tad

Tad was a boy who loved to have fun. He found schoolwork hard and did not learn to read until he was 12.

Some people saw the president as a plain country lawyer. They were wrong. He may have begun that way, but he was a learner, and he kept changing and growing. Abraham Lincoln was a complicated man: gifted, sophisticated, ambitious, shrewd, kindly, humorous, thoughtful, and very intelligent.

He had a rare sense of compassion. That means he not only cared deeply about others, but he suffered with each hurt they suffered. So as the war progressed he got sadder and sadder. Sometimes he had bad dreams, awful forebodings, and dark depressions.

But there was something that could always cheer him up. It was the antics of his two younger sons, Willie and Tad. They were merry boys, often mischievous, who turned the White House into a playground. Sometimes, in the middle of an important meeting, one or the other would come into Lincoln's office, climb in his lap, give him a hug, and then disappear.

Tad was really Thomas, but he wiggled so much his father nicknamed him Tadpole. One day he fired his toy cannon at members of the cabinet. Another time he was playing in the attic and discovered the controls of the White House bell system. If someone wanted a member of the White House staff he or she pulled a cord and a bell rang. Tad, of course, pulled all the bells and the place went wild. Was there a fire? An emergency? What was going on? When they discovered Tad, he was on the attic floor, laughing away. Another time he sneaked into the White House kitchen and ate up all the strawberries intended for a fancy dinner.

Tad and Lincoln played games together in the White House. When they played blind man's buff, the president would trip over the furniture on purpose so that Tad could escape.

Willie was quieter. He wrote poetry and liked to read, but he did his share of giggling, running through the White House corridors, and playing pranks. He had a sweet manner, and his father was especially proud when a poem the 10-year-old wrote was published in the newspaper. Those who knew him said he was the boy who was most like his father.

Both Tad and Willie loved animals, and their parents let them turn the White House into their own kind of zoo. They had ponies, a turkey, white rabbits, kittens, a pet goat, and a dog named Jip. The boys liked to play soldier and were fascinated with the war. Lincoln often took them with him when he went to visit nearby troops. Sometimes he left them at home—or thought he did. Once, when the president was at an encampment talking to some soldiers, he saw a mule wagon heading his way. The mule drivers turned out to be Willie and Tad, and they were holding wooden swords high above their heads. Sometimes they invited their friends to parade around the president's mansion while they banged drums and blew on horns.

Now, as you might guess, this disturbed some of the distinguished visitors. The politicians and generals and cabinet members complained and said the boys were spoiled and should be spanked. But the president and his wife, Mary Todd Lincoln, wouldn't consider it. Lincoln called them his "two little codgers" and often played right along with them.

There was an older son, Robert, who was more serious and off at Harvard as a student. A fourth son, Eddie, had died at age four when they lived in Springfield, Illinois. The Lincolns never got over that loss; perhaps that was why they seemed to love their boys with such abandon.

Robert Lincoln became a lawyer and served in Chester Arthur's government as secretary of war.

Mary Todd Lincoln had a hard time as First Lady—for some she could do nothing right. One visitor said, "She stuns me with her low-necked dresses and the flower beds which she carries on top of her head."

When Willie died (this is a mourning portrait of him, with a lock of his hair), Lincoln said, "I know that he is much better off in heaven, but then we loved him so. It is hard, hard to have him die!"

When 12-year-old Willie Lincoln died, his brother Tad was nine. Elizabeth Keckley, a friend of the family, wrote of Lincoln, "I never saw a man so bowed down with grief."

Mary Lincoln had a hard time being the president's wife. When she came to Washington people laughed at her as well as at her lawyer husband. When they laughed at him, he told a joke and laughed back. But Mary couldn't do it. Criticism hurt her. People called her a hayseed and acted as if she were some kind of bumpkin. That wasn't true at all. Her father had been a state senator; her family was prominent in Kentucky. She was witty, she loved to talk, she often entertained, and she had a husband who loved her. She would show those Easterners. She would have a party that was fancier and finer than any before in Washington.

And so she decorated the White House, bought a beautiful dress, and sent for celebrated caterers to prepare an elegant meal. Then the boys got sick, and the president and his wife thought of canceling the party. But the invitations had been sent and the doctors didn't seem worried.

Tad got better, but Willie's fever wouldn't go away. He got weaker and weaker. The party was grand, as fine as Washington had seen. But all night long the president and his wife took turns going upstairs and sitting with Willie and putting wet towels on his head.

Although they didn't know it, Willie had typhoid fever, probably caught from the contaminated water that was a result of the overcrowding and carelessness and poor sanitary practices of wartime.

When he died—well, it is too sad to tell the details. The president and his wife had suffered a war death of their own. It was almost more than they could bear.

This cartoon was drawn at the beginning of 1863, a few months after Willie's death. The Union was doing badly in the war, Lincoln was deeply sad about Willie, and the outlook was grim. The cartoonist predicted that more government heads were going to roll.

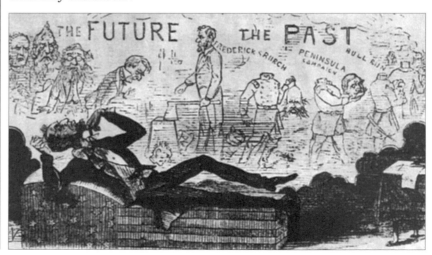

18 General McClellan's Campaign

Lincoln (right) occasionally managed to visit General McClellan (far left) on the battlefield, but mostly he wrote impatient letters urging the general to do something and attack the enemy.

After Manassas (Bull Run) the North got serious. General George B. McClellan was in charge, and he was a man who could organize an army. "I found no army to command," he said, "just a mere collection of regiments cowering on the banks of the Potomac."

He brought order and pride to those regiments. He drilled the young recruits and saw that they knew how to fight and march and camp. He made sure they had enough food and equipment. He organized an army of 100,000 men—more than had ever been commanded by one man in all the history of the western hemisphere.

By spring of 1862, that army was ready to fight. McClellan had a brilliant plan. He planned to capture Richmond. Richmond was not only the capital of the Confederacy, it was one of the few industrial cities in the South. Guns and steel and even uniforms were made there. If Richmond fell, the Confederacy might collapse and the war would be over.

McClellan's plan was to take most of the army by boat to the Virginia Peninsula—that land between the York and James Rivers. Look at the map (page 91) and at the peninsula. Do you see some familiar places? How about Yorktown and Williamsburg? Now think about moving an army from Washington, D.C., to the bottom of that Virginia Peninsula.

Abraham Lincoln to Gen. George McClellan: *It is indispensable to you that you strike a blow…you must act.*
General McClellan to his wife: *I was much tempted to reply that he had better come and do it himself.*

Large herds of cattle marched with the army to be slaughtered for food.

The U.S. Army, camped at Cumberland Landing on the Pamunkey River during the Peninsular Campaign, waiting for marching orders.

I don't see the sense of piling up earth to keep us apart. If we don't get at each other some time, when will the war end? My plan would be to quit ditching and to go to fighting.
—UNION PRIVATE J. W. REID

This cartoon sneered at McClellan for directing the troops from the safety of a gunboat, away from the action. One of Lincoln's letters to him said, "Dear General, if you do not want to use the army I would like to borrow it for a few days."

McClellan had 100,000 men, 2,500 supply wagons, 300 cannons, and 25,000 animals to worry about. Four hundred ships were kept busy transporting and supplying the Union forces. The army used 600 tons of supplies each day.

According to the plan, once the Union army was in place it was to march north, up the peninsula, to Richmond. There it was to meet another Union army marching south from Washington. It was that anaconda, squeeze-the-enemy strategy. It was a fine plan, but a few things went wrong.

First, there was McClellan himself. He was great at planning and organizing, but very slow and cautious when it came to action. He got his army landed and organized on the southern tip of the Virginia Peninsula, and then he just sat there and organized some more. That gave the Confederate army time to get ready.

One thing McClellan didn't plan on was rain. That spring of 1862, there was more rain than normal in Virginia. The peninsula turned into a place of mud and ooze. The soldiers joked, "Virginia used to be in the Union. Now it's in the mud."

Mud makes marching difficult. Especially if you are marching on dirt roads with cannons, horses, mules, and herds of cattle.

The Union army got way behind schedule. The Confederates used that time well. Stonewall Jackson and his men were fighting Union soldiers in Virginia's Shenandoah valley. They headed

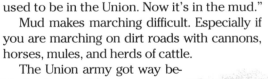

toward Washington. Now there was cause for alarm: the nation's capital was in danger. The army that was supposed to march south and meet McClellan stayed in Washington.

Meanwhile, the Confederates on the peninsula were using their wits. John Magruder, an officer who was an amateur actor, knew a lot about dramatic presentation. He marched a small force in and out of the woods in a continuing circle. A band played loud music and the soldiers marched noisily. Edmund Patterson from Alabama wrote home that he had "been travelling most of the day, seeming with no other view than to show ourselves to the enemy at as many different points as possible. I am pretty tired." The ruse worked. McClellan was sure he faced an enormous force.

General John Magruder

He had the Union soldiers march slowly and cautiously. But, finally, McClellan's army was at the outskirts of Richmond; they were six miles from the Confederate capital. Union soldiers could hear the city's church bells. Now McClellan seemed really frightened. He believed the Confederate army was much larger than the Union army. (Actually, the opposite was true.) He hesitated. The Confederates took advantage of McClellan's indecision. Robert E. Lee was a man who rarely hesitated. He attacked. For seven days the two armies battled fiercely. Both sides fought bravely. Both sides suffered terrible losses. In the end, there was no clear winner.

But McClellan's confidence was destroyed. He would not stay around to continue the fight. He ordered his army to retreat. He ran away from the fight. All those months of planning, transporting, marching, and

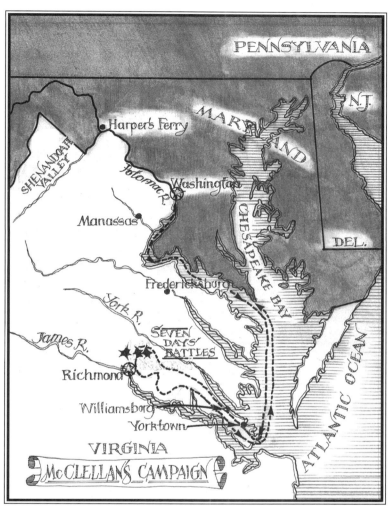

PENNSYLVANIA

N.J.

Harper's Ferry

MARYLAND

SHENANDOAH VALLEY

Potomac R.

Washington

CHESAPEAKE BAY

DEL.

Manassas

Fredericksburg

ATLANTIC OCEAN

York R.

SEVEN DAYS' BATTLES

James R.

Richmond

Williamsburg

Yorktown

VIRGINIA

McCLELLAN'S CAMPAIGN

Now our eyes saw [that] five thousand dead or wounded men were on the ground.... enough of them were alive and moving to give the field a singular crawling effect.

—COL. WILLIAM W. AVERELL

There was a full moon at the battle of Chancellorsville, but it had not yet risen when Jackson was shot. "Wild fire that, sir," said the general. "Wild fire."

fighting, and nothing gained. McClellan's complicated, expensive war plan had failed to accomplish much of anything. The Union army went back North. The Yankees had lost the chance to capture Richmond and end the war. It would be two years before it came again.

The Rebels hadn't accomplished much either in those seven days of awful fighting. They had had a few chances to finish off the Yankee army and they hadn't done it.

The next two years brought many bloody encounters and much frustration—and not just on battlefields in the East. Terrible battles were fought in Tennessee and Mississippi and Louisiana and Missouri. In the West the Union forces often seemed to be ahead, in the East it was the Confederacy—but mostly it was a draw.

But something did happen in that time that made a difference. It was at a battle near Chancellorsville, Virginia, in the spring of 1863, in an area so dense and thick with trees that people called it the Wilderness. Lee was outnumbered almost two to one, and yet he won the fight. Many say it was his most brilliant victory. But it wasn't worth the cost. Stonewall Jackson was wounded by one of his own men in the smoke and confusion of battle. (In war it is not unusual for soldiers to be hit by bullets fired by their own companions.)

After Jackson's left arm was amputated he seemed to be getting better; then he got pneumonia, which often happened after battlefield injuries. General Lee sent a message: "Tell him to make haste and get well, and come back to me as soon as he can. He has lost his left arm, but I have lost my right arm."

The message pleased Jackson, and he told it to his wife, Anna. Stonewall had never talked much about his private life, but he loved Anna and was proud of their new baby, Julia. They had come to be with him, but they couldn't help him. Jackson was in terrible pain and was often delirious. He was fighting his final earthly battle when he

called out, "Prepare for action," and "Let us cross the river and rest under the shade of the trees." And then he died.

In Richmond, 20,000 hushed, tearful people lined the streets as four white horses pulled his coffin in a solemn military cortège. When his body was brought home to the Virginia Military Institute, no one talked of Tom Fool. He was now a hero, and would remain so.

In a book named *Stonewall,* author Jean Fritz writes: *Years later at a reunion of Confederate veterans in Richmond, a group of gray-haired men were found sleeping on the ground around Stonewall's statue in Capitol Square. They were among the few survivors of the old Stonewall Brigade. "We were his boys," one veteran explained, "and we wanted to sleep with the Old Man just once more."*

19 War at Sea

The *Cairo* was a Union ironclad and the first ship to be sunk by the torpedo—now called a *mine*—which was perfected by the Confederates. One inventor, Matthew Maury, worked on the torpedo in his bathtub at home in Richmond.

Right away—after the first shells were fired at Fort Sumter—Lincoln ordered the Union navy to blockade Southern ports. That was a tough order. The U.S. Navy had only 90 ships, and that included ships on the Atlantic Coast, the Gulf of Mexico, the Mississippi, and on other inland rivers.

The Confederacy had no navy at all. Both sides got busy. Northern shipyards worked at top speed. By the time the war ended, the North had about 700 ships in service. The Confederates knew they could never match the Yankees when it came to numbers of ships, so they secretly ordered fast cruisers from English shipyards. (England was neutral and not supposed to help one side or the other. The shipbuilders pretended the ships were going to Italy.) The sleek, agile cruisers were designed to be able to attack Northern cargo ships. The most famous Confederate cruiser, the *Alabama*, captured 62 Yankee merchant ships in two years before it was finally sunk.

The Southerners did something else. They decided to coat some of their ships with iron. Up to that time almost all ships were made of wood. A few ironclad ships had been experimented with in Europe—but none had been tested in battle. Wooden ships were fast and easy to float; they were also fragile, and flammable. Throw a match on one and—*whoosh!* A ship with a layer of iron all over it would be difficult to set on fire. An iron ship could also be a powerful weapon: imagine it ramming into a wooden boat.

A Union vessel named *Merrimack* had sunk near Norfolk, Virginia. The Confederates raised the ship out of the water, cut down her

> There was a craft such as the eyes of a seaman never looked upon before—an immense shingle floating on the water, with a gigantic cheese box rising from its center; no sails, no wheels, no smokestack, no guns. What could it be? [It was the *Monitor.*]
>
> —LT. JAMES H. ROCHELLE, CSS *PATRICK HENRY*

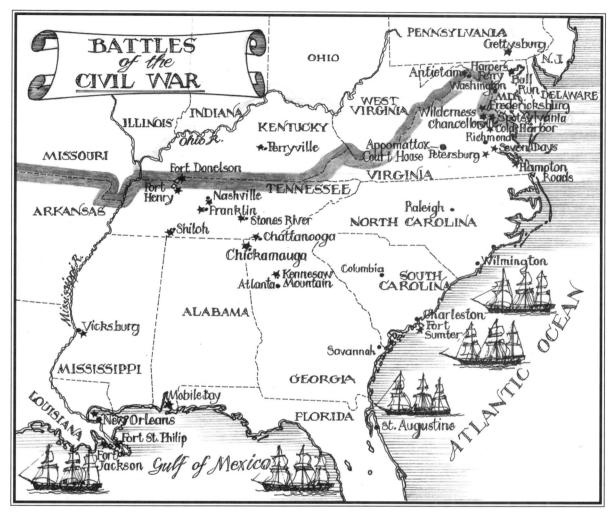

BATTLES of the CIVIL WAR

burned top, and covered her with slanting metal plates. That produced a weird-looking boat; actually, it looked just like a floating roof. On each side of the roof were holes for five powerful guns. A center chimney let out steam from the engines. The *Merrimack* was big and clumsy and moved very slowly, but that metal coat made it strong and fierce-looking. Now that she had a new form she needed a new name. The Confederates called her the *Virginia*.

It had taken months to get the *Virginia* ready but, finally, on March 8, 1862 she slid into the water for what everyone thought was a trial run. Everyone but the men in charge. They intended to fight that day, and fight they did. Before the day was over two big Union frigates were

Raphael Semmes, captain of the Confederate ship *Alabama*. His exploits make exciting reading.

95

Above, officers of the U.S.S. *Monitor* in a picture taken not long after the battle with the *Virginia* (shown below). On December 30 of that year, 1862, the *Monitor* sank in a storm off Cape Hatteras. (It was found in 1973, lying in 220 feet of water.) The *Virginia* returned to Norfolk, its home port. A couple of months later McClellan captured Norfolk and the *Virginia* was burned.

destroyed and dozens of men were dead. Three Union ships came hurrying to the rescue—they would have been destroyed, too, but their captains soon realized what was going on and turned around and fled. Shells that hit the *Virginia* bounced harmlessly off her sides.

When news of the battle reached Washington, people were frantic. Lincoln's cabinet was meeting and the secretary of war, Edwin M. Stanton, was beside himself. You can imagine the discussion: the South had unleashed a new superweapon. The secretary of war said the *Virginia* could sink every vessel in the North; it could steam up the Potomac River to Washington and "disperse Congress, destroy the Capitol and public buildings." (That wasn't true; fear sometimes makes people think hysterically.)

But there was something that even Secretary Stanton did not know. Long before the launching, spies had informed the navy department that the *Virginia* was being built. So the U.S. Navy decided to build its own ironclad vessel. That ship, the *Monitor*, was on its way to meet the *Virginia* at Hampton Roads. (Hampton Roads is the Virginia waterway where three rivers come together, meet Chesapeake Bay, and flow into the ocean. The watery "road" is six miles wide—from the Hampton shore to the Norfolk shore.)

The *Monitor* was as strange-looking as the *Virginia*. Strange in a different

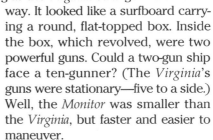

way. It looked like a surfboard carrying a round, flat-topped box. Inside the box, which revolved, were two powerful guns. Could a two-gun ship face a ten-gunner? (The *Virginia*'s guns were stationary—five to a side.) Well, the *Monitor* was smaller than the *Virginia*, but faster and easier to maneuver.

It was Sunday, March 9, 1862, when they met. The shores of Hampton Roads were lined with people. No battle like this had ever been fought before. If the *Monitor* lost, the Union-held Fortress Monroe, on the Hampton side, could be starved into surrender.

For four hours they battled. The *Virginia*'s guns kept firing, but the *Monitor* was an impossible target—

Ruler of the President's Navy

When David Glasgow Farragut was a boy he was adopted by David Porter (who was a naval hero in the War of 1812). He wanted to be like his adopted father, so he went to sea and was a midshipman by the time he was 9 years old. At 13 he was in command of a captured British ship. Farragut was 58 when the Civil War began; he volunteered to fight for the Union cause. That was surprising, because he was a Southerner from Alabama, but Farragut believed in the United States.

In 1862, Farragut was the commander of a fleet of 17 ships sent to blockade the mouth of the Mississippi River. He set out to capture

"Damn the torpedoes! Full steam ahead!" said Farragut in 1864 at the battle of Mobile Bay.

New Orleans, the South's largest city and busiest port. To get to the city Admiral Farragut had to pass two heavily armed Confederate forts. In addition, a chain of hulks (old ships) was laid across the Mississippi to keep Yankee ships out. "Nothing afloat could pass the forts," said a citizen of New Orleans. Farragut's fleet headed up the river. The Confederates opened fire. One ship was hit 42 times. Farragut's flagship was set ablaze. "The passing of the forts…was one of the most awful sights and events I ever saw or expect to experience…[it] seemed as if all the artillery of heaven were playing upon the earth," said Farragut. That didn't stop him. His sailors put out the fires and got all the ships moving. They did what they intended.

To weaken the outer defenses of New Orleans before moving up the river, Farragut bombarded the Confederate garrisons at Fort Jackson and Fort St. Philip for five terrible days. Then he went 70 miles upriver to the city, smashing the Confederate fleet on the way.

They captured New Orleans. A few months later Farragut sailed his ships up the river to Vicksburg, through more Confederate fire. That gave the North control of most of the Mississippi. (By July 1863, the Union had the whole river.) Farragut wasn't finished; he had more exploits and heroism ahead of him. Abraham Lincoln said David Farragut was his best appointment of the war.

it was small and agile, and never stayed where it was expected. It didn't matter; neither ship was able to sink the other. Cannonballs bounced off their sides. The battle was a draw, although both sides claimed victory. Actually, it was iron ships that won; that battle finished wooden warships.

Charles Francis Adams, who was U.S. minister to England (and John Quincy Adams's son), wrote from England that the battle of the *Monitor* and the *Merrimack* "has been the main talk of the town even in Parliament, in the clubs, in the city….The impression is that it dates the commencement of a new era in warfare."

Northerners never used that name, Virginia. They called the ironclad by its original name, Merrimack.

20 Emancipation Means Freedom

A soldier's grave under a tree at Antietam. (Union soldiers usually named battles after nearby water —Antietam, Bull Run, Pittsburg Landing; Rebels after the town—Sharpsburg, Manassas, Shiloh.)

Lincoln needed a victory because he wanted to do something important. He wanted to make an announcement, he wanted to change the purpose of the war, and he didn't want to do it as the leader of a discouraged, defeated army.

Then finally it came. It wasn't the kind of victory Lincoln had hoped for. Too many men had been killed; more than ever before. But the Northern army had stopped General Lee's army at Antietam Creek, and that would have to do.

The battle of Antietam is sometimes called the battle of Sharpsburg, because Antietam Creek is near the little town of Sharpsburg, Maryland. McClellan had some unexpected luck just before the battle at Sharpsburg. One of his men found three cigars wrapped in some paper. The cigars probably dropped from the pocket of a Confederate officer. The paper showed Robert E. Lee's battle plans. "Here is a paper," boasted McClellan to an old West Point friend, "with which if I cannot whip Bobbie Lee, I will be willing to go home." McClellan now knew exactly where all of Lee's forces would be. Even with that help, he didn't whip Robert E. Lee on September 17, 1862—the bloodiest day of the war— but he did stop him.

Both sides suffered terrible losses at Antietam Creek, but when it was over things were worse for the Confederates. They were hungry and exhausted, they had few supplies, and they were in easy reach of

Two weeks after Antietam Lincoln visited McClellan on the battlefield. One soldier said that he "seemed much worn and distressed."

the Yankee army. Another attack by the North and they might have been finished. But two days after the battle, McClellan watched the Rebels retreat. He let them cross the Potomac River, back to Virginia.

Lincoln took a train to Sharpsburg and urged McClellan to go after Lee. McClellan had two divisions of fresh soldiers. "I came back thinking he would move at once," said Lincoln. "I...ordered him to advance." McClellan sat for 19 days; then he moved slowly. He let the Confederate army get away. Lincoln was enraged. He believed he had lost a chance to end the war. He wasn't the only one. Northern soldier Elisha Rhodes wrote a letter home saying, "Oh, why did we not attack them and drive them into the river? I do not understand these things. But then I am only a boy."

(Soon after Antietam Lincoln did send McClellan home; McClellan did not go willingly. He said, "They have made a great mistake. Alas for my poor country!")

Still, Antietam was a victory for the North, and an important one. Robert E. Lee had been on his way to the northern state of Pennsylvania. He had intended to cut important railroad lines in Harrisburg, Pennsylvania. He had hoped to find shoes and supplies for his army in the North. Now a discouraged Confederate army was back in Virginia. Lincoln could make the announcement he had planned for several months.

Lincoln now changed the war from a

Angel of the Battlefield

Clara Barton was a clerk in the Patent Office in Washington when war broke out. She was a tiny woman, but full of energy, and she used it visiting homesick troops from her native Massachusetts. She brought soap and lemons and homemade food. But when she saw the suffering of the soldiers after the battle of Bull Run, she knew she could do more. So she organized an agency to get supplies for wounded soldiers. But that wasn't enough for Clara Barton; she decided to go where she was needed, where the fighting was happening. At the battle of Antietam, she arrived with a wagonload of bandages, anesthetics, and oil lanterns. Then she set to work in a field hospital: bandaging, feeding, and consoling the wounded. When shells began exploding close by, most of the male nurses ran for cover. Barton stayed and held the operating table steady for a surgeon who called her "the true heroine of the age, the angel of the battlefield." After the war was over, Clara Barton founded the American Red Cross.

When the war was over Barton led a search for missing soldiers.

This anti-Lincoln cartoon shows him writing the Emancipation Proclamation surrounded by devilish pictures and objects.

fight to save the Union into something much greater. He changed it into a battle for human freedom, a battle to end slavery. He did that with a document called an *Emancipation Proclamation*. It said that all the slaves in the rebel states were free. I'm going to repeat that: *all the slaves in the Rebel states were free.*

It was about time, said the abolitionists. It was the right time, said Lincoln.

He read the proclamation in September of 1862; it became official on January 1, 1863. It didn't free slaves in the North. (There were no slaves there to free.) It didn't free slaves in the border states. (The president didn't have the power to do that. It could only be done by a constitutional amendment, or by the state legislatures.). It freed slaves in the Confederate states. (Where Lincoln had no power.)

Today, when you think about that, it doesn't sound as if Lincoln really did much. People in 1862 knew differently. They knew what he had done was very important. When he signed that document it meant there would be no going back. When the war was over there would be no chance of compromise on slavery. Slavery was dead in the South. And, if it was dead there, it would soon die in the border states.

People's ideas had changed. And Lincoln's ideas, too. He had never liked slavery—in fact he hated it—but he thought it more important to save the Union than to end slavery. He had come to realize that was impossible. Slavery was like a worm in a good apple—it was making the whole apple rotten. It was no longer enough just to save the Union. Who wants to fight for a nation with a rotten core? The nation could not allow an evil practice and believe in itself. It could not speak of liberty and equality and be cruel and hateful to a large group of its own people.

And so the Civil War became a revolution. Not a revolution as the Southern leaders saw it. They still thought they were fighting for

We shout for joy that we live to record this righteous decree...."free forever"...oh, ye millions of free and loyal men who have earnestly sought to free your bleeding country from the dreadful ravages of revolution and anarchy, lift up now your voices with joy and thanksgiving for with freedom to the slave will come peace and safety to your country.

—FREDERICK DOUGLASS

independence. What those leaders wanted was independence to oppress other people.

No, this was a revolution in men's and women's ideas. They had been led by the logic of Thomas Jefferson's Declaration of Independence. Was this a nation that really believed that all people have the right to "life, liberty and the pursuit of happiness"? And that all "are created equal?" Or were those just empty words that applied only to special people? If blacks could be enslaved, then what about Northern factory workers, or war prisoners, or any group of people?

The war was making people think. It was making them question ideas they had never really considered. Are some people actually better than others just because their skin is a different color? Perhaps green-eyed people are the best, or short people. Could curly hair be superior to straight hair? And what about baldies? Maybe they would make good slaves.

One thing everyone knew: this war, begun with such gaiety, had become a national nightmare. It was not a game. The corpses of American boys were being shoveled into the earth—hundreds of thousands of them. And so people asked themselves: *What is this war really about?*

It was clear in the South. Southerners were fighting to preserve their way of life. It was a way of life that was often elegant and attractive, but it was based on a mean, despicable practice—slavery. If the South was fighting to preserve slavery, then the North must be fighting to end it. And so it was.

But it took Abraham Lincoln and most Northerners some time to figure that out. Perhaps that was because there was much racial prejudice in the North, and that had to be overcome.

Voices were speaking out against that racial hatred; finally people began listening to them.

One of the voices belonged to a man who spoke clearly and said the North must "lift the war into the dignity of a war for progress and civilization."

The man's name was Frederick Douglass, and he had been a slave and had whip marks on his back to prove it. Douglass had learned to read and write—against great odds—and was one of the most eloquent writers and speakers of his time. He knew it was folly to fight just to save the Union. Douglass wrote:

Of Thee I Sing

On January 1, 1863, Thomas Wentworth Higginson stood before a group that included black soldiers, white soldiers, civilians, and slaves. In a simple ceremony he raised the American flag and announced that the slaves were now free. "There suddenly arose," wrote Higginson, "a strong male voice (but cracked and elderly), into which two women's voices instantly blended, singing, as if by an impulse that could no more be repressed than the morning note of the song sparrow—'My Country, 'tis of thee, Sweet Land of Liberty, of thee I sing!'"

Do you think Southern slave owners would be likely to free their slaves because their enemy said they should do so?

John Tenniel, who drew this picture of Lincoln as a gambler playing his last card—the Emancipation Proclamation—was also the artist who made the illustrations for a famous book: *Alice in Wonderland.*

To fight against slaveholders, without fighting against slavery, is but a half-hearted business, and paralyzes the hands engaged in it. Fire must be met with water....War for the destruction of liberty must be met with war for the destruction of slavery.

And thus the Civil War became a war to make the United States what it had meant to be from its beginnings—a fair nation, a great nation. A nation that fulfilled the best ideas of its founders; a nation that would set equality of opportunity as a goal; a nation that could promise "life, liberty and the pursuit of happiness" and mean it for all its peoples.

Usually the president signed government bills with a simple *A. Lincoln*. But when he signed the Emancipation Proclamation, Abraham Lincoln wrote his name in full. "Gentlemen," he said to the cabinet officers standing near him, "I never, in my life, felt more certain that I was doing right than in signing this paper."

Abroad, the proclamation attracted friendly attention as well as dislike. A French newspaper ran this engraving of *Negroes Celebrating the Emancipation Proclamation*.

21 Determined Soldiers

This sergeant served in Texas. The noncommissioned rank of sergeant was as high as a black soldier could rise.

What would you think if you learned that a large group of strong men who wanted to fight for the Union were turned down? Would you think that was dumb? You're right. It was worse than dumb —it was idiotic. But racial prejudice is always like that—it is always stupid.

Prejudice turns up in all times and places, but in the 19th century it was a sickness that infected much of the nation—North as well as South.

If the South had been without prejudice there would have been no war; if the North had been without prejudice the war might have been much shorter. There were large numbers of blacks who wanted to fight in the war. Because of racial prejudice the Union army wouldn't have them—at least not at first.

Now this may sound ridiculous to you, but many people actually believed that blacks couldn't fight. That is what they believed and said. Many whites had forgotten—or never knew— that blacks had been explorers and mountain men and early settlers. African Americans fought in the Revolutionary War and the War of 1812. Of course they wanted to fight in this war. But neither side would have them as soldiers—not the North or the South.

In the South slaves helped the war effort; they had no choice. They raised crops and did farm work so that white men were free to be soldiers. Slaves worked in Southern factories. (Yes, the South did have ironworks and factories—just not as many as the North had.) Rebel armies used slaves to build and cook and do other necessary work. The Confederacy

What upon earth is the matter with the American people?...The national edifice is on fire. Every man who can carry a bucket of water ...is wanted....[Yet government leaders] refuse to receive the very class of men which has a deeper interest in the defeat and humiliation of the rebels than all others....Such is the pride, the stupid prejudice and folly that rules the hour.

—FREDERICK DOUGLASS, BLACK ABOLITIONIST LEADER

On the left is Taylor, and on the right is the drummer boy he became in the 78th Colored Troops in Louisiana. That is all we know about him.

103

Escaped slaves (contrabands) on a Union army wharf in Virginia. At first the South said runaways had to be returned under the Fugitive Slave Law; but the North replied that the Confederacy had declared itself a foreign country, and U.S. laws no longer applied there.

It is not too much to say that if this Massachusetts 54th had faltered when its trial came, two hundred thousand troops for whom it was a pioneer would never have been put into the field....But it did not falter. It made Fort Wagner such a name for the colored race as Bunker Hill has been for ninety years to the white Yankees.

—THE NEW YORK TRIBUNE

could not have fought long without slaves.

When they could, African Americans ran away to Yankee army camps. At first the Union officers didn't know what to do with them. A few sent the slaves back to slave masters, but most let them stay. The Union officers called the blacks *contrabands*. Contraband of war is property seized from the enemy—especially property that can help the war effort. Soon the contrabands were doing useful work for the Northern armies.

In July 1863, at Ft. Wagner, S.C., nearly half the all-black 54th Massachusetts died. When the flag bearer fell, Sgt. William Carney (right) saved the colors despite several bullet wounds. Sgt. Carney was awarded the Congressional Medal of Honor.

But what they really wanted to do was to fight.

They wanted to fight because they cared about America as much as anyone else. They wanted to fight because they knew that fighting men would never be thought of as slaves again.

They wanted to fight because they knew—long before most white people—that this was a war about slavery.

Finally it became possible. A month after Lincoln read the Emancipation Proclamation, a group of black contrabands fought as soldiers in Missouri. Ten died, the first black combat victims of the war. Soon there were legions of black soldiers. They fought well. The assistant secretary of war visited General Grant's army and said that

> *the bravery of the blacks completely revolutionized the sentiment of the army....I heard prominent officers who formerly in private had sneered at the idea of negroes fighting express themselves after that as heartily in favor of it.*

Massachusetts organized a regiment of black soldiers (called the 54th Massachusetts) under the command of a young (white) Bostonian, Colonel Robert Gould Shaw. The men of the 54th Massachusetts led a bayonet attack on Fort Wagner—a massive fort of wood and earth that stood on an island at the entry to Charleston Harbor. The Confederates, protected by the thick walls of the fort, shot cannons and naval guns at the charging men. Almost half of them were wounded, captured, or killed. After that no one asked if blacks could fight. "Prejudice is down," wrote a man who was there.

Everyone knew it took special courage for blacks to fight and for white officers to fight with them. They realized that, if they were captured, they would not be treated like the other soldiers.

Confederate officials had announced that any captured white Northern officers who led black troops would be put to death as criminals. Captured black soldiers knew they would be sold into slavery, if they weren't murdered. Nonetheless, they kept volunteering. Before the war was over, more than 180,000 black soldiers fought with the Union army. Lincoln said they made victory possible.

In 1864 a newspaper reporter wrote about black troops marching in New York City. He said:

> *Had any man predicted it last year he would have been thought a fool....*

You say you will not fight to free Negroes. Some of them seem willing to fight for you. [When victory is won] there will be some black men who can remember that, with silent tongue and clenched teeth, and steady eye and well-poised bayonet, they have helped mankind on this great consummation. —ABRAHAM LINCOLN

These men of the 107th Colored Infantry helped to defend Washington, D.C.

You cannot make soldiers of slaves....The day you make soldiers of them is the beginning of the end of the revolution. And if slaves seem good soldiers, then our whole theory of slavery is wrong.
—SENATOR HOWELL COBB, GEORGIA

At Fort Pillow, Confederate troops massacred 300 of the fort's mostly black defenders.

THE MASSACRE AT FORT PILLOW

Official Confirmation of the Report.

Three Hundred Black Soldiers Murdered After Surrender.

Fifty-three White Soldiers Killed and One Hundred Wounded.

RETALIATION TO BE MADE

Eight months ago the African race in this City were literally hunted down like wild beasts [in the New York race riots]. They fled for their lives....How astonishingly has all this been changed. The same men...now march in solid platoons, with shouldered muskets, slung knapsacks, and buckled cartridge boxes down through our gayest avenues...to the pealing strains of martial music and are everywhere saluted with waving handkerchiefs, with descending flowers.

The *New York Times* reporter said that the marching soldiers were a sign of the *marvelous times*; he also said there was a *revolution [of] the public mind*. What did he mean by that? And what did Abraham Lincoln mean when he said, *In giving freedom to the slave, we assure freedom to the free*?

In 1862 Robert Smalls (inset) was a slave pilot on the Confederate steamer *Planter* (below). One night the captain was ashore. Smalls sailed the ship out of Charleston and turned it over to the Union blockading squadron. At right, black sharpshooters practice in camp.

22 Marching Soldiers

"I felt strange enough, lying down this my first night in camp," one young man wrote in his journal. But soon a soldier found he could sleep almost anywhere, any time.

I can tell you something for sure: you wouldn't want an army marching through your front yard! Picture tens of thousands of soldiers on the march. Where do they sleep? What do they eat? How do they stay warm and dry?

The first thing Civil War soldiers did when they made a new camp was to cut down trees and fences. Sometimes they built huts for shelter. Even when they slept in tents, they often cut wooden supports to make the tents stronger and more comfortable. Sometimes they pushed people out of their houses and used the houses. To keep warm they burned logs, furniture, books, or whatever else they could find.

The army quartermaster supplied them with flour, coffee, bacon, and a heavy cracker called *hardtack*. But most soldiers wanted fresh meat and vegetables. So they took food from the farms they passed; they trampled fields where they marched.

As the war went on, the generals began to realize that anything they could do to weaken the enemy

Confederate troops usually had to improvise their winter quarters, since they didn't have the Union army's tents and other supplies. These men from the First Texas regiment in Virginia built a log cabin for their mess and named it for General P. G. T. Beauregard, who captured Fort Sumter.

would help win the war. If they made life difficult for ordinary people in their homes, those people might stop supporting the war. So the armies began destroying barns and crops on purpose. It was a strategy called *total war*, and it brought grief to the civilian population.

Mostly it was the South that suffered, because most of the war was fought on Southern land. Much of it in Robert E. Lee's beloved Virginia. The South was being ruined. General Lee wanted the North to suffer. He knew there were strong peace movements in the North. If his soldiers won some Northern territory, if they beat the Yankees on their own home ground, he thought the North would soon beg for peace.

He might have been right. It was an enormous gamble, but Lee's daring had made him successful in the past. He decided to go for it. He decided to march his army north.

Remember, General Lee had marched north before, and had been

Men of the 159th New York shopped at the U.S. Hotel D' Grub for "dride appels," "rice pudin," and "meels at all ours." But there was "no licker sold to soljers"—or so the sign said.

stopped at Sharpsburg, Maryland—by Antietam Creek. That was in September of 1862. But Antietam was really a draw. The Confederates had been stopped, but not beaten. After Antietam, Lee had two spectacular victories in Virginia: one at Fredericksburg, the other at Chancellorsville. Now, in the early summer of 1863, the Confederates were confident. The South seemed to be winning the war. Robert E. Lee decided to take the war north: he headed for Pennsylvania.

The U.S. Army went after them. It had a new leader—again. This time the army was commanded by thin, tough, hot-tempered George G. Meade. He didn't seem to be afraid of Robert E. Lee, which was unusual for a Union general.

These Yankee soldiers have taken over a plantation in Hilton Head, South Carolina.

The two armies met in a quiet college town named Gettysburg, where roads come together like the spokes of a wheel. The town lies in a gentle valley with hills and ridges and knolls about. In 1863, Gettysburg was twelve blocks long, six blocks wide, and had 2,400 inhabitants. A spectacular battle had been fought there earlier: millions and millions of years earlier. Gettysburg is the very place where the continents split apart in dinosaur days. Rock that was once near Gettysburg went off in different directions. Some of it now lies in Iceland. No one knew that in 1863, but anyone who had been to Gettysburg knew there was a lot of very hard stone in the area. Just south of the town, huge boulders sprawled one on top of another in a weird

Some of that Gettysburg rock is one billion years old.

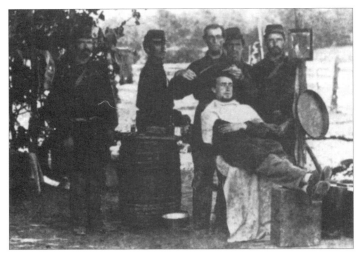

Getting a trim and a shave. Today the military has strict rules about long hair and beards, but in the Civil War the most important reason for cutting hair was to keep lice at bay.

Ricochet (RICK-uh-shay) means to bounce off something, like a billiard ball off the edge of a pool table.

jumble that looked like a giant's playground. Gettysburg people called the area Devil's Den. Those boulders were so hard bullets would ricochet off them.

Devil's Den was part of a fishhook-shaped ridge of land that ran from two rock-littered hills, past the Den, past a wheatfield, past a peach orchard, to a hill—Cemetery Hill—covered with tombstones.

It was postcard-pretty land: green and peaceful. A wooden sign on Cemetery Hill said anyone firing a gun in the area would be fined five dollars.

If, on the first three days of July in 1863, the town could have collected those five-dollar fines, it would have been the richest place in America. No question: it was the bloodiest.

The little town of Gettysburg, Pennsylvania, in 1863, the year it became the scene of one of the worst battles in American history.

23 Awesome Fighting

General George Pickett led the most famous charge in the Civil War (maybe in U.S. history).

No one planned to fight in Gettysburg. But when a Rebel officer saw an advertisement for a shoe store it made him head for that Pennsylvania town. He was searching for shoes for his soldiers when he ran into Union cavalry. That started things.

Actually there weren't shoes—or much of anything—left in Gettysburg. When the townsfolk heard that soldiers were in the area they took their cattle and sheep and store goods and got them out of town. Gates Fahnestock's dad and uncles owned the biggest store in Gettysburg. The store was called Fahnestock Brothers. The Fahnestocks rented a railroad car, filled it three times, and sent almost everything in the store off to Philadelphia for safekeeping.

Gates, who was 10, was more interested in soldiers than in store matters. He climbed up on the roof of his house to watch the Rebel soldiers march into town. They came shouting and firing guns into the air, but Gates couldn't help noticing that many of them were barefoot and in rags.

Like almost everyone else in Gettysburg, Gates didn't cheer those Southerners. He wanted the Union to win. So when General John Buford's U.S. cavalry rode into town he did cheer. Then a bullet whistled over the roof and he hightailed it inside.

Eighteen-year-old Daniel Skelly was perched in a tree on a nearby ridge. He and a group of men and boys wanted to see some action. But

The flowers in bloom upon the graves at the Cemetery were shot away. Tombs and monuments were knocked to pieces, and ordinary gravestones shattered in rows.

—PRIVATE WARREN GOSS

The Fahnestock store at the corner of Baltimore and Middle streets. After the battle it was commandeered for the local headquarters of the U.S. Sanitary Commission, an organization that helped the sick and wounded—something like the Red Cross (which Clara Barton hadn't founded yet).

John Burns—70 years old and a War of 1812 veteran—was the only Gettysburg civilian to fight in the battle. He got tired of watching, picked up his old flintlock musket, and reported for duty to the 150th Pennsylvania on the afternoon of the first day. He was wounded three times but survived.

The blood stood in puddles in some places on the rocks.
—Colonel Wm. C. Oates

A Massachusetts artillery battery held its position at this farmhouse near the peach orchard for a few desperate hours. After they were forced to retreat they left 50 horses lying dead. If you visit Gettysburg today you can still see a cannonball hole in the barn.

when "shot and shell began to fly over our heads, there was a general stampede towards town."

It was July 1, 1863, and the fireworks were about to begin.

Neither army had all its soldiers in place—many were marching toward Gettysburg—but that didn't stop these armies. They were eager to fight. They wanted to finish this war; it had gone on too long. The soldiers who marched into Gettysburg singing were soon fighting like savages. First one side seemed to be winning, then the other. By afternoon it looked as if it would be another Union defeat.

Rebels had pushed the Union forces through the town. When they stormed into Gates Fahnestock's house they discovered eight Union soldiers; one was hiding in the potato bin. They asked Gates to get some hay. He did, and they piled it up in front of his house and made comfortable beds for themselves.

The Rebels had won the first day's battle, but fresh soldiers were pouring into the area. General Buford sent a message to General Meade. "Get soldiers here at once. This is a good location for a fight."

Buford's horsemen had discovered that Gettysburg sits on a series of ridges with a shallow valley in between. Whoever held the high ground would have a big advantage. The Yankees had fewer troops than the Rebels, but Buford and his men were on Cemetery Ridge, south of the village. It was a good place to be. Any attack by the Confederates would have to come uphill.

The Confederates had a problem. Their dashing, show-off cavalry star, J. E. B. Stuart (who was called the laughing banjo player), had gone off in the wrong direction. No one knew where he and his men were. Cavalry are the eyes of the army. The Southern army was blind. The Confederates had no idea of the topography of the area. Still, they were confident. After the big win at Chancellorsville it was hard to imagine that Union soldiers could beat them.

The night of July 1 was warm and clear and the sky was full of stars. Union soldiers slept in the cemetery and prayed they would see another night.

Confederates were saying the same prayers in their camp on Seminary Ridge, another high area. A religious school—a Lutheran seminary—was located on that ridge. The two ridges, about a mile apart, were separated by woods, open farmland, and peach orchards.

July 2 came, and the fighting began early and was even more murderous than on the 1st—and that was bad enough. That the day was blazing hot didn't make a difference. Men fought and died with a frenzy that is still hard to believe. A Northern regiment—the 1st Minnesota—had 262 men when the day began; 42 survived. By nightfall the land was cluttered with the bodies of men and horses. When that terrible second day ended each side thought itself ahead, although the deaths made both sides losers.

And then July 3 dawned, and the battle that some say decided the war (although it would be a while before that was understood). Let's watch what happens. General Lee now has fewer men than General Meade. That doesn't scare Lee; he has been outnumbered before. At Chancellorsville he fought and beat an army twice the size of his.

The Confederates have a bold plan. They begin by blasting the Union line with a cannonade fiercer than anything tried before: two hours of nonstop artillery fire. "The very earth shook beneath our feet," wrote a soldier, "and the hills and rocks seemed to reel like drunken men." Gates Fahnestock and his family huddle in their cellar. The noise and the shaking of the house are like an earthquake that doesn't stop.

Confederate Major General George E. Pickett has arrived in Gettysburg with fresh troops. Pickett is known to be courageous. His soldiers are ready to reinforce the Rebel line and Pickett is eager to fight. He will lead the most famous military charge in all of American history.

But General Longstreet doesn't like it a bit. He is Robert E. Lee's most trusted aide. Theirs is an unusual friendship. Longstreet is big, rumpled, tough-talking, and sometimes crude. He says what he thinks. Lee is a courtly gentleman, always polite, always in control.

Lee has planned a great military charge across three-fourths of a mile

[The rebel lines were] at once enveloped in a dense cloud of dust. Arms, heads, blankets, guns and knapsacks were tossed into the clear air. A moan went up from the field.
—A UNION OFFICER

Gen. Longstreet disagreed with Lee's tactics at Gettysburg. After the war he became a Republican and worked for President Grant.

At times I saw around me more of the enemy than of my own men: gaps opening, wallowing, closing again; squads of stalwart men who had cut their way through us, disappearing….All around, a strange, mingled roar.
—JOSHUA LAWRENCE CHAMBERLAIN, 20TH MAINE

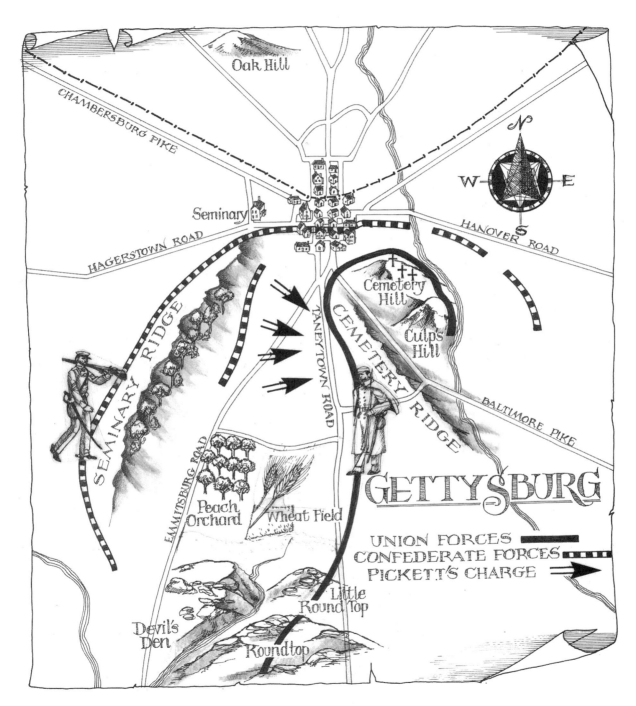

Oak Hill

CHAMBERSBURG PIKE

Seminary

HAGERSTOWN ROAD

HANOVER ROAD

N
W — E
S

Cemetery Hill

Culps Hill

CEMETERY RIDGE

BALTIMORE PIKE

SEMINARY RIDGE

TANEYTOWN ROAD

EMMITSBURG ROAD

Peach Orchard

Wheat Field

GETTYSBURG

UNION FORCES
CONFEDERATE FORCES
PICKETT'S CHARGE

Little Round Top

Devil's Den

Roundtop

of open fields. Longstreet argues against it. He wants to pull out of Gettysburg. He believes in defensive fighting. But Longstreet loses the argument. Defensive fighting isn't heroic to most soldiers. The Confederate army prepares to advance toward Cemetery Hill.

Suddenly it is quiet. The shelling has stopped. Line after line of gray-uniformed men march out of the woods. The bright sun reflects on silver bayonets and red flags. George Pickett, riding a white horse, "his jaunty cap raked well over his right ear, and his long auburn locks…hanging almost to his shoulder," shouts his orders. Then, elbow to elbow, like a grand parade, 15,000 awesome fighters step forward in an incredible, orderly, moving rectangle almost a mile wide and half a mile deep.

> *Right on they move, as with one soul, in perfect order, over ridge and slope, through orchard and meadow and cornfield, magnificent, grim, irresistible.*

Lee has gambled everything on this old-fashioned military charge. It is like something from the Middle Ages—from the days of knighthood. The charge is glorious, and daring. If it works, who knows? The heart may go out of the Yankees.

Winfield Scott Hancock is in command of the Yankees who are crouched behind a low stone wall on Cemetery Ridge. For two hours Rebel shells have been exploding over their heads: the noise, the July heat, the smoke, and the tension of waiting for battle have been almost more than the men can bear. Luckily for them, the cannons are aimed poorly, and much of the shot has gone too high and fallen behind the lines. It has hit ambulances, and men eating in the grass, and men in tents. But it has not done the harm intended. The Confederates don't know this. They march on cheerfully. Later, Confederate Captain W. W. Wood will remember, "we believed the battle was practically over, and we had nothing to do but march unopposed to Cemetery [Hill] and occupy it."

On that hill, the Union soldiers are having a hard time staying calm, especially now, as the massed Rebel army advances. It is a sight these men will never forget.

The Confederate *artillery barrage began at one o'clock on July 3.*

The storm broke upon us so suddenly that numbers of soldiers and officers who leaped from their tents or lazy siestas on the grass were stricken in their rising with mortal wounds and died some with cigars between their teeth.

—A Union soldier

Pickett's charge. A Massachusetts soldier said, "Foot to foot, body to body, and man to man they struggled and pushed and strived and killed. The mass of wounded and heaps of dead entangled…hatless, coatless, drowned in sweat, black with powder, red with blood."

After being wounded, Hancock sent a message to General Meade: "I did not leave the field until the victory was entirely secured. The enemy must be short of ammunition as I was shot with a ten-penny nail."

Man touching man, rank pressing rank...the red flags wave, their horsemen gallop up and down; the arms of thirteen thousand men, barrel and bayonet, gleam in the sun, a sloping forest of flashing steel.

"It was," said a U.S. colonel, "the most beautiful thing I ever saw."

Hancock gives the order for the Union artillery to fire. The cannons fire more than cannonballs. The new fire-power includes cans that explode and send off a hail of metal that murders the oncoming soldiers. Bravely they close ranks and keep coming.

This famous picture of a dead sniper in Devil's Den was set up. The photographer dragged the body 40 yards and arranged it dramatically.

The commander says, "Let them come up close before you fire, and then aim low and steadily."

Imagine: you are there. Your gun is resting on the stone wall. You're a sharpshooter, and you are using a rifled musket that is accurate at half a mile, but even if you were a bad shot it would be hard to miss this target.

The Confederates, a tight mass of men in light gray uniforms, are caught in deadly fire. The Yankees are shooting at them from different angles all along Cemetery Ridge. This is a massacre. Some Rebels run away, but most keep coming, stepping over the bodies of those who have fallen before them.

A few make it to the stone wall. One who does—an officer—puts his hat on the tip of his sword, holds the sword high—and charges. He leaps the low stone wall and is followed by 200 soldiers. They don't get far. The brave officer is Confederate general Lewis Armistead, General Hancock's old friend. Armistead is hit. A Union officer rushes to his side, awed by the valiant Confederate. Armistead asks the officer to give General Hancock his watch; he sends Mrs. Hancock his Bible. They are his last wishes. Hancock, too, has been wounded, but he will recover.

The charge is over. The men around you sing "John Brown's Body" and the "Battle Hymn of the Republic." You cheer the victory, and then you weep for your dead friends.

In the Rebel camp, General Lee, great leader that he is, blames no one but himself. He has used a straight down-the-center frontal attack. That strategy is out of date. The generals—on both sides—don't seem to understand that. It will be tried again.

24 Lee the Fox

After Gettysburg, Lee (sitting) asked Jefferson Davis to relieve him of command, but the Confederate president refused. Then, said Lee, "We must now prepare for harder blows and harder work."

Lee needs to get his tired men home. The Confederates have lost (killed, wounded, or missing) 28,000 men: that's one third of their army. There is no more talk of invading the North. It has begun to rain—hard—and the Potomac River, which they must cross, is flooding. Wagons carrying the 5,000 wounded men stretch for 17 miles. Many will die before they reach home. Many who rode north now walk south; thousands of horses have been killed.

Lincoln sees a chance to finish Lee's army. Lee is trapped, his men are exhausted. The president expects action. GENERAL MEADE ATTACK, YOU HAVE FRESH SOLDIERS IN YOUR CAMP. But once again Lincoln has a general who is cautious. Besides, Meade is facing a general known as the Gray Fox. Lee, the fox, sends a soldier into the Northern camp. The soldier pretends to be a Rebel deserter; he tells the Yankees the Confederates are ready to fight. Meade hesitates, the river goes down, and the Confederate army makes it back to Virginia. "We had them within our grasp," moans Lincoln. "We had only to stretch forth our hands and they were ours." But they didn't grasp. The war will continue.

Still, when news of the victory at Gettysburg arrives in Washington

Union losses at Gettysburg are 23,000, one fourth of their army.

"If General Meade [above] can complete his work," said President Lincoln, "the rebellion will be over." But Meade couldn't.

The people of Vicksburg could not stay in their houses during the siege; it was too dangerous. They camped in bomb shelters dug out of the city's steep hillsides.

We are utterly cut off from the world, surrounded by a circle of fire. The fiery shower of shells goes on, day and night: People do nothing but eat what they can get, sleep when they can, and dodge the shells.

—DORA MILLER, VICKSBURG

on the Fourth of July, the city celebrates. After two glum years, many think it the most glorious Independence Day since that first one in 1776. In a few days people will learn that on this same July Fourth General U. S. Grant has won a major victory in Mississippi.

Like Stonewall Jackson, Ulysses Grant is a leader who has only one goal: to win. He will let nothing stand in the way of that goal. And, again like Jackson, he knows when and how to break the rules of custom.

In Mississippi, Grant moved his army where no one thought an army could be moved. He left his supply base (no general is supposed to do that); he decided to take a risk. He believed he could feed his men with food from the farms in the Mississippi countryside. He was right. Grant's army laid siege to Vicksburg. The navy blockaded the river entry to the city. Vicksburg's citizens were trapped. They couldn't leave, and no one could help them. Then Union cannons began shelling the city. The siege lasted 48 days. Before it was over, those in Vicksburg were eating rats—and anything else they could find. Grant starved and bombed them into surrender. The 30,000 Confederate soldiers in the city surrendered, too.

Union forces now control Vicksburg, and that means they control the Mississippi River. Abe Lincoln can say of the river, "The father of waters flows unvexed to the sea." (What does he mean by that?)

Those two victories—Gettysburg and Vicksburg—reverse the war. Gettysburg shows that Robert E. Lee is beatable. Vicksburg is a great strategic victory. Later, Grant says, "The fate of the Confederacy was sealed when Vicksburg fell."

But the Confederate leaders are stubborn. So are the Confederate soldiers. They will not give up; they will not return to the Union. As the South becomes desperate, Southern soldiers fight harder than ever.

Do you remember when Cortés faced Moctezuma in Mexico City? Do you remember that the Aztec leader would not surrender? Do you remember what happened?

25 Speeches at Gettysburg

One reason Lincoln worked hard on his speeches was that he did not speak well extemporaneously. In his election campaign he made off-the-cuff speeches that gave him political trouble.

Weeks after the battle at Gettysburg, bodies still lie on the ground unburied. Like almost everyone else in town, Gates Fahnestock goes to Cemetery Hill to help with the sorrowful work. But when Gates sees the bodies and smells the smells he throws up. A soldier gives him a piece of rag to tie over his nose, like a bandit's mask. Gates cuts wood and fetches water for those who are doing the burying.

The citizens of the little town of Gettysburg are overwhelmed by the tragedy around them. In addition to the dead, some 16,000 wounded soldiers have turned every house, barn, and building into a hospital. A Confederate general is lodged in young Charles McCurdy's barn. "One day, in the privacy of the stable, he took off his shirt and showed me his back on which a full rigged ship was tattooed." Charles thought it a thrilling sight.

When five-year-old Willie Cunningham dies, his parents blame disease caught from a sick soldier. Tourists and sightseers and grieving families are now pouring into Gettysburg. Some come to help. Eighteen states agree to share the costs of a national cemetery to be established on Cemetery Hill. The dead soldiers—all Yankees—will finally rest in peace.

A ceremony is planned for November 19 to honor them. Edward Everett is to be the main speaker. He is a fine choice. Everett has been president of Harvard, a senator, secretary of state, and ambassador to Britain. He is thought to be the greatest orator of the day; all are sure

In 1848 Edward Everett was president of Harvard. Students protested when a black applied to the college. Everett said, "If this boy passes the examinations he will be admitted; and if the white students choose to withdraw, all the income of the college will be devoted to his education."

On November 20, 1863—the day after the ceremony at Gettysburg—13-year-old Allen Frazer found a shell left on the battlefield. When he picked it up it exploded; he was killed instantly.

he will say the right things.

A few weeks before the occasion someone thinks to invite the president. No one expects him to accept the invitation. But he does. He is asked to make "a few appropriate remarks." It is not intended that he say much—it is Edward Everett people will come to hear.

It takes six hours by train from Washington to the Pennsylvania town of Gettysburg. Lincoln doesn't want to miss this occasion; he comes a day early and then settles in his room to work on his speech. It is already written, but he revises and rewrites—as is his way. (You can see the speech in Lincoln's handwriting in the Library of Congress in Washington, D.C.)

He will use this opportunity to try to explain the meaning of the war. Many Northerners are crying out for peace. They no longer care about the Union, or the slaves. In New York, recently, there were riots when calls came for more soldiers. Lincoln knows the nation can have peace any time it wants. But that would end the United States. It would end the Founders' dream: the dream that people can govern themselves; the dream that people are "created equal" with inalienable rights to "life, liberty, and the pursuit of happiness." Lincoln believes this terrible war has a purpose. He believes it must give a new birth to the dream.

Thousands of men and women have gathered at Gettysburg. They include famous citizens, veterans, cabinet members, and ordinary people. All march together to the battlefield. A band plays and the marchers form a parade. Abraham Lincoln rides a brown horse. Someone has chosen poorly. The horse is tiny, and Lincoln's legs almost touch the ground. Along the way a mother hands her daughter to the president. He puts the little girl in front of him and she rides too. Someday she will tell her children and grandchildren how she rode with the president at Gettysburg.

The president with his secretaries, John Nicolay (left) and John Hay. Hay, who kept an interesting diary of his time in the White House, said of Lincoln, "It is absurd to call him a modest man. No great man is ever modest."

The distinguished guests climb onto a wooden platform and wait for Edward Everett before the ceremonies can begin. Everett is getting ready in a special tent set aside for his use.

He talks for almost two hours, without notes, in a voice deep and rich. Later, no one seems to remember what he said, but they knew he said it well. There are prayers and other speeches, and the 15,000 listeners who sit or stand in the afternoon sun are hot and tired when the president finally rises, puts on his steel-rimmed glasses, and reads his few remarks. His is a country voice, and it sounds dull after the polished tones of orator Everett. A snicker runs through the crowd.

The speech takes two minutes. When the president finishes there is not a sound—not a clap, not a cheer. People seem to have expected more.

The speech was a failure; so Lincoln tells his friends, and so he believes. But he is wrong. The presidency has changed Lincoln. He has grown in greatness. He has learned to use words as a poet uses them—with great care and precision. He had been an able country lawyer with a good mind and a taste for jokes. Now he is much more than that. The deaths and the burdens of war are making him noble, and thoughtful, and understanding, and sad.

Most of those who listen that day at Gettysburg do not know they are hearing one of the greatest speeches ever written, but all know their president is speaking from his heart. A few understand that this is no ordinary speech. Edward Everett says, "Mr. President, I should be glad if I could flatter myself that I came as near to the central idea of the occasion in two hours, as you did in two minutes."

This is what the president said on November 19, 1863; this is how he explained the meaning of the war:

Four score and seven years ago our fathers brought forth on this continent, a new nation, conceived in Liberty, and dedicated

When the draft was announced in New York City, many people protested. Blacks were attacked because the rioters saw them as the reason for the war—and for the draft, too.

A score is 20 years. Lincoln wanted to remind Americans of the number of years since 1776. He wanted to remind them of the Founding Fathers and their purposes. He could have said the number (what is it?), but *four score and seven* has a solemn, majestic sound. He liked the music of those words.

121

The crowds and marching bands at Gettysburg. Below is the first page of the address, in Lincoln's own handwriting.

to the proposition that all men are created equal. Now we are engaged in a great civil war, testing whether that nation, or any nation so conceived and so dedicated, can long endure. We are met on a great battle-field of that war. We have come to dedicate a portion of that field, as a final resting place for those who here gave their lives that that nation might live. It is altogether fitting and proper that we should do this.

*B*ut, in a larger sense, we can not dedicate— we can not consecrate—we can not hallow this ground. The brave men, living and dead, who struggled here, have consecrated it, far above our poor power to add or detract. The world will little note, nor long remember, what

There was no official photographer at the Gettysburg Address. A local man took some pictures, and Lincoln is just about visible here (in square). His speech was over so fast that the slow cameras of the time wouldn't have managed it anyway.

we say here, but it can never forget what they did here. It is for us the living, rather, to be dedicated here to the unfinished work which they who fought here have thus far so nobly advanced. It is rather for us to be here dedicated to the great task remaining before us—that from these honored dead we take increased devotion to that cause for which they gave the last full measure of devotion—that we here highly resolve that these dead shall not have died in vain—that this nation, under God, shall have a new birth of freedom—and that government of the people, by the people, for the people, shall not perish from the earth.

If you can only remember one thing from this book, make it the Gettysburg Address. Go ahead and read it aloud. Then read it again. And again. You'll be surprised, it won't be long before you know it by heart.

Why is this civil war a test of our nation? Can a nation dedicated to liberty and equality last? Did these deaths have a purpose? What is the reason for this ceremony?

What work is unfinished? What do you think Lincoln means by a *new birth of freedom*? What happens if the war is lost? What is the challenge to the people of America? Is it still a challenge?

26 More Battles— Will It Ever End?

General Grant at army headquarters with his wife, Julia, and his youngest son, Jesse. Grant was much happier with his family. Alone, he got depressed and drank too much.

Finally, Abraham Lincoln found the strong general he'd been looking for.

He asked Ulysses S. Grant to take charge of all the Union armies. Lincoln had noticed that nothing seemed to scare Grant. But then nothing seemed to scare Robert E. Lee either. Those two warriors—Grant and Lee—were now pitted against each other. Lee was an aristocrat: dignified—and tough. Grant was an everyday kind of fellow: soft-spoken—and tough.

Both were devoted family men, and decent, and their followers respected that.

Grant was anxious to fight it out with Lee and end the war. Both he and Lee knew that the longer the fighting continued the more likely it was that northern people would get tired of supporting the war and give up. The Southern strategy was to wear the North out; the Northern was to end things as quickly as possible.

Grant led a huge army south from Washington—he was heading for Richmond. He had almost 120,000 men. Lee had a little more than half that number.

Grant attacked, and attacked, and attacked. One dreadful battle, called the battle of the Wilderness, was fought in thick woods. It was

Union and Confederate cavalry in the Wilderness. The North wasn't defeated, but it had so many killed or wounded (18,000) that it didn't win either.

horrible. The trees caught on fire, and no one could see through the trees and smoke. Tens of thousands of men were killed or wounded or burned in that blind inferno.

Neither general was a quitter, nor was either the kind of leader who conserved men. The death lists became gruesomely long. The Confederates won the battles—they lost fewer men than the Yankees—but the losses hurt them more: they had fewer men to lose. And Grant kept pushing on, getting closer and closer to Richmond.

Grant soon learned what every other Union general knew—you couldn't attack Robert E. Lee and win. Grant had to trap the Confederate army if he was to be victorious. He made plans to lay a siege (as he had at Vicksburg).

The WILDERNESS to PETERSBURG

UNION ADVANCE · CONFEDERATE ADVANCE · CONFEDERATE FORTIFICATIONS

If you check the map you'll see that Petersburg and Richmond are side-by-side cities. Splendid cities, and unusual. Both combined the industrial spirit of the North with the graciousness of the South. Petersburg was the supply center for Lee's army. Five railroad lines came into Petersburg, so did several major roads. Two rivers—the James and the Appomattox—meet at nearby City Point. Petersburg was a key. Without food from there, Richmond (the Confederate capital) would starve.

Grant decided to besiege Petersburg. But first he had to get there, and that wouldn't be easy. Petersburg is on the southwest side of the broad James River. Grant and his army were northeast of the James. Lee's army was in between (in and around Richmond) and not about to stand aside and let 100,000 men march by.

It is hard to believe what happened—but it was something amazing. Grant and his army disappeared. For three days the Confederate army couldn't find them. The Northern army was marching south—quickly and quietly. The Confederate army lost track of them, and that gave the U.S. Army engineers just enough time. They put pontoons in the river.

> The conduct of the Southern people appears many times truly noble as exemplified, for instance, in the defense of Petersburg; old men with silver locks lay dead in the trenches side by side with mere boys of thirteen or fourteen. It almost makes one sorry to have to fight against people who show such devotion for their homes and their country.
> —WASHINGTON ROEBLING, JULY 7, 1864

The Army of the Potomac crossing the Rapidan River by pontoon bridge on its way to the Wilderness. The horses on the nearer bridge are part of an artillery battery, pulling a heavy gun behind them.

Picture rowboat-like pontoons, one next to another, stretched across the water. Then imagine wooden planks connecting the pontoons, and a road laid on those planks. Now see in your mind an army marching across that bridge. An army that stretches back for 50 miles!

The bridge was 10 feet wide and almost half a mile long (from one side of the river to the other). When men, wagons, horses, and artillery marched on it, the bridge went up and down and up and down. The James River has a swift-moving current, so the bridge also swayed from side to side. It took two days to get the whole army across the river. Then the pontoon bridge was broken up and towed away. The U.S. Army could now march up the west side of the James to City Point. Grant was ready to begin his siege of Petersburg and Richmond. It went on for 10 long, dreary months. Ten months with some fighting and shelling, but mostly it was waiting time: boring, nerve-racking, and tense.

Are you getting tired reading about this war? Well, imagine that you live in Petersburg and the siege is on. You are hungry—very hungry. Some citizens have siege parties to keep up their spirits. At one party the only refreshment is water from the muddy James River.

Now imagine that you have a brother or father fighting in the war. Or maybe you are a soldier—a starving Confederate stuck in Petersburg all through a long, hot summer and then a cold, gloomy winter. Try being a Yankee soldier. You're not hungry, but you are weary of living in a tent, being far from your family, and never knowing when a shell may head your way.

A fair number of Rebels just sneak off from the camps. Many are mountain boys who never cared about slavery; they are tired of this war. They run away from the army; they desert. The penalty for desertion is death. So the deserters have a problem. They can't go home. They can't get caught. Bands of them roamed the South, living as outlaws. They have to steal to live. They make life difficult for the southern people.

By 1864, Southerners didn't need extra troubles; they were having a hard enough time. All of the fighting was now in the South. It was destroying the country. The armies were gathering food from the farms. When they couldn't use the food themselves they sometimes burned it so the other side wouldn't get it.

War is the remedy our enemies have chosen...and I say let us give them all they want.
—WILLIAM TECUMSEH SHERMAN

General Sherman's army marched on Atlanta in two huge columns several miles apart (see map on next page). They cut down forests to build corduroy roads; they pulled up railroad tracks—twisting them into "Sherman's hairpins" and "Jeff Davis neckties"—to make sure the Confederates couldn't use them again. When they got to Atlanta, they found the city surrounded by trenches and fences—built by stripping the walls of nearby houses.

Then General William Tecumseh Sherman marched a big army —the U.S. Army of the West— from Tennessee, to Georgia, and on to the Carolinas. That army's march is one of the most famous in military history.

Red-bearded General Sherman was a friend of U. S. Grant. Before the war he had been head of a military academy in Louisiana. He was asked to be a Confederate officer. But, although he had many Southern friends, he believed in the Union. A soldier described Sherman as "tall, spare and sinewy, with a very long neck, and a big head....a very homely man, with a regular nest of wrinkles in his face, which play and twist as he eagerly talks on each subject....his expression is pleasant and kindly." Sherman's men called him "Uncle Billy."

Uncle Billy was about to squeeze the South like that anaconda snake. Like Grant at Vicksburg, Sherman broke the rules he had learned when he was a cadet at West Point. When he marched east he left his supply lines, and an army is never supposed to do that. Sherman gambled that he could find enough food in the agriculturally rich South. He was right. But the Yankees didn't have it easy. They

Our own conviction is...that it is advisable for the Union Party to nominate for President some other among its able and true men than Abraham Lincoln.
—HORACE GREELEY, NORTHERN NEWSPAPER EDITOR, MAY 13, 1864

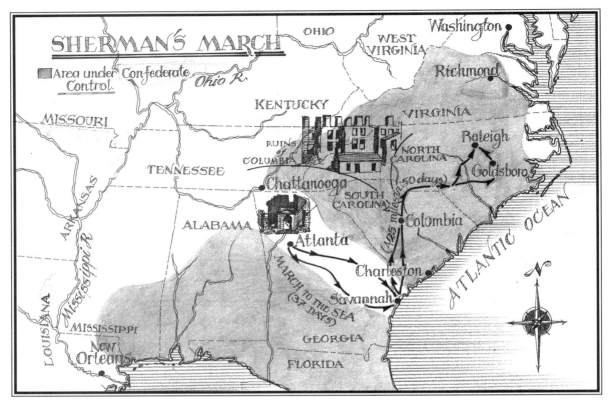

SHERMAN'S MARCH

Area under Confederate Control.

"Is that you, still there, Long Abe?" moans Jefferson Davis. "Yes! And I'm going to be four years longer," says Lincoln, confirming Davis's nightmare that he has been reelected.

fought a Rebel army much of the way.

Before long the Northern soldiers were out of control. They, too, were tired of the war and they blamed the Southerners for it. So they stole and burned and destroyed the country as they went south. Sherman's army cut a path 40 miles wide. Nothing much was left on that path but empty fields, burned barns and homes, and slaughtered livestock.

Sherman believed that army battles alone would not win the war. He believed in "total war." He thought that the Southern people who were supporting the war needed to be hurt. He believed that the Southern ability to make war—its food and arms production—must be destroyed. His army did a lot of destroying. It was cruel, because many innocent people were hurt. But Sherman may have been right. He probably shortened the war.

For a while, up North, no one knew what he was doing. General Sherman seemed to be lost somewhere in Georgia. There were rumors that he was losing battles. People in the North were sick of the war. It wasn't on their doorsteps, but in every town people were mourning their lost sons. The person they blamed was the president.

They could do something about him. Lincoln was running for reelection. General George McClellan was running against him. McClellan called himself a peace candidate. He said he would end the war quickly. He didn't think slavery should be allowed in the western territories, but he had no intention of prohibiting slavery in the South. His supporters called him a great general who had been misunderstood by the president. His opponents said he meant to the end the Union. A leading Southern newspaper said McClellan's election as president would "lead to peace and our own independence."

Everyone was tired of the war. Many no longer cared about the Union. Some didn't care about slavery. It looked as if Lincoln would lose. He was sure of it. "I am going to be beaten," he said, "and unless some great change takes place, badly beaten." He might have been right, but a great change did take place. General Sherman won an astonishing victory: he captured the city of Atlanta.

The ruins of Columbia, South Carolina, after Sherman's army had passed through.

Atlanta is ours and fairly won.
—GENERAL SHERMAN

The Turn of the Tide

There were no telephones during the Civil War. That was lucky for future historians. You see, when Mary Jones wanted to say something to her son, Charles Colcock Jones, Jr., she had to write a letter. In her letters she told him all about the things that were going on at home and how she felt about the war. Charles, an officer in the Confederate army, told her what he was doing and what was on his mind. Thousands and thousands and thousands of letters like theirs were written during the Civil War.

And thus we know that Abraham Lincoln was right when he thought he might not win re-election as president. We can read

letters and see that. The people in the North were calling the president a "Lincolnpoop," and worse. Then, suddenly, the tone of the letters began to change.

Military victories seem to have made the difference. An army in northern Virginia, led by General Philip Sheridan, beat a Confederate army and destroyed southern farmland. The crops from that land—in the Shenandoah Valley— were meant to feed

The destruction of Shenandoah Valley farmland during "Sheridan's Ride" was known there as "The Burning."

the Confederate army. When Sheridan finished, there was so little grain left it was said that a crow flying overhead had to carry its own lunchbox. This, and Sherman's capture of Atlanta, made people realize war's end might be near. President Lincoln began to look good to them again.

27 The Second Inaugural

Lincoln the giant beat McClellan the shrimp in the election. McClellan's spade recalls his days as general; it was said his army did more digging than fighting.

Abraham Lincoln won reelection. By a big margin. The American people were beginning to appreciate the tall man who was their president. They were beginning to understand his greatness: to understand that he was kind, compassionate, and humble. And also to understand that he was strong and determined.

Four years earlier, when he was first elected, he had hoped to prevent war. He had been willing to do almost anything to keep the country from splitting apart. He had even been willing to allow slavery to continue in the South. Now he felt differently. The war had been ghastly, much worse than anyone had ever imagined it could be; but now he saw a purpose in that war. Slavery would be ended. It might not have happened without the war. He thought it was all part of God's mysterious way, and that was what he said in his second inaugural address, the speech he gave after being reelected president.

The war was winding down. Sherman was devastating the Deep South; Grant was laying siege to Petersburg; troops were threatening Richmond. Lincoln knew the South could not hold out much longer. But what would become of the nation when the war was over? How should the defeated South be treated? Were they enemy? Or were they brothers and sisters who had behaved badly?

Suppose you have a terrible fight with a friend. Suppose that friend really hurts you. What would you do when the fight is over? Hate the friend? Or shake hands and say, "Let's forget it"?

Sometimes it isn't easy to forgive a former enemy. People in the North were hurt. Many wanted to punish the South. Lincoln thought the

That month and a half, from Lincoln's inauguration on March 4, to April 15, is an important time in American history. Hold on, things will begin to happen very fast.

war had been punishment enough. He wanted to welcome Southerners back into the United States—as long as they agreed to play by the rules of the Constitution and the Emancipation Proclamation.

Rebuilding the country was going to be difficult. Rebuilding the South would be especially difficult. He was worried about that. That process of rebuilding was being called *Reconstruction*. Lincoln wanted Reconstruction to be carried out in a friendly, helpful way. Slavery had been wrong, terribly wrong, but now it was done with. In his second inaugural address he said it was not a just God's purpose to have men "wringing their bread from the sweat of other men's faces."

Wringing their bread from the sweat of other men's faces. Lincoln was talking to those who approved of slavery when he used that phrase. (What did he mean?)

He was talking to those who hated the slave owners when he used a thought from the Bible and said: *Let us judge not that we be not judged.* (What does that mean?)

In that famous second inaugural address, Abraham Lincoln told the American people—North and South—that he wanted them to be kind and generous toward each other. He used these eloquent words:

After the election, *Harper's* drew a very long Abe. At the inaugural ball, Mary Lincoln (below, left) wore a silk dress that cost $2,000—a fortune in those days.

Needing Nevada!

Nevada's residents weren't in a hurry for Nevada to be a state, but Lincoln was in a rush. He made sure the territory became a state in 1864, one week before his second election. He needed Nevada's three electoral votes in the presidential race, and its two votes to get the slavery-ending 13th Amendment passed.

Before the election, Lincoln toted up the electoral votes he had to win, adding Nevada to be certain.

In the Gospel according to St. Matthew of the New Testament, Jesus says, "Judge not, that ye be not judged."

I have been up to see the [Confederate] Congress and they do not seem to be able to do anything except eat peanuts and chew tobacco, while my army is starving.

—ROBERT E. LEE

Lincoln's second inauguration. When he heard the election results, he said, "I give thanks to the Almighty for this evidence of the people's resolution to stand by free government and the rights of humanity."

With malice toward none, with charity for all; with firmness in the right, as God gives us to see the right, let us strive on to finish the work we are in; to bind up the nation's wounds; to care for him who shall have borne the battle, and for his widow, and his orphan—to do all which may achieve and cherish a just, and lasting peace, among ourselves, and with all nations.

Charles Francis Adams, Jr., who was John Adams's great-grandson, said of the president's reelection speech: "That rail-splitting lawyer is one of the wonders of the day....Not a prince or minister in all Europe could have risen to such an equality with the occasion."

Lincoln was pleased with the speech too. When a journalist complimented him, the president said, "I expect the [second inaugural] to wear as well as—perhaps better than—anything I have produced."

28 Closing In on the End

"His hair was every which way for Sunday," Henry Ward Beecher said of Lincoln. "It looked as though it was an abandoned stubble field."

I daily part with my raiment for food. We find no one who will exchange eatables for Confederate money. So we are devouring our clothes.
—MARY CHESNUT

If I had another face, do you think I'd wear this one?
—ABRAHAM LINCOLN

Grant (standing, wearing hat) and his staff at headquarters in City Point. Seated (far right) is engineer Ely Parker. You'll read about him in the next chapter.

"I'm a tired man," said Lincoln to a visitor. "Sometimes I think I am the tiredest man on earth."

He had reason to be tired. There were all those cares and pains of four years of war, and Willie's death; that fight for reelection, and a battle to get Congress to pass the 13th Amendment to the Constitution. (The 13th Amendment said there would be no slavery in the United States. It went further than the Emancipation Proclamation; it made emancipation the law of the whole land.) Besides all that, there was a never-ending parade of visitors in the president's office. Some were people who wanted jobs, some were congressmen and cabinet members with problems to discuss, some were newsmen, some were just busybodies. Lincoln tried to see them all.

On March 14, 1865—10 days after his second inauguration—he was so tired he held a cabinet meeting in his bedroom. He was in bed, propped up with pillows. His secretaries and those who knew him best were alarmed. Maybe a change would help. He needed to get away from Washington and that long, tiring daily river of visitors.

Perhaps Grant heard of his fatigue. He sent the president a message by telegraph wire: *Can you not visit City Point for a day or two? I would like very much to see you, and I*

Above, the *River Queen* outside City Point, where Lincoln docked for two weeks. Below, the crucial battle of Five Forks, at which General Philip Sheridan swept down on the Confederate infantry, capturing four cannons and several thousand soldiers.

think the rest would do you good.

A few days later, on March 23, Lincoln and his wife and young son Tad were aboard the yacht *River Queen* on their way to Grant's headquarters at City Point. They stayed two weeks.

It was pleasant aboard ship off City Point. Oh, it did rain some, but the river breezes were mild and the air was full of the fragrance of new spring growth. From the presidential yacht, Lincoln looked out at the military might of the Northern army. The high, jutting piece of land that is City Point was filled with officers' cabins and tents. A comfortable old plantation house was military headquarters. Long, strong docks had been built so supplies could be unloaded easily. The Yankees brought trains and laid railroad tracks to take supplies the 13 miles from City Point to Petersburg. Northern engineers worked hard and with much efficiency. Because of that efficiency, Northern soldiers had plenty of food and equipment during the long siege.

City Point may have been the busiest place in the nation that April. Lincoln watched as ships and people kept streaming in. He was pleased to see one of the other visitors: it was General Sherman. Sherman had left his army in North Carolina to come and consult with General Grant. Lincoln talked of peace with the two military leaders. All agreed: the war had been harsh but the peace should be gentle.

The three of them knew that Lee and his army would soon be starved out of Petersburg and that the Confederate government would soon be starved out of Richmond. Would the Rebels then surrender? Or would they choose to fight on?

The Confederate leaders did not consider the choice. Many of them feared they would be hanged as traitors if captured. Many had such a strong belief in their cause they could not imagine surrendering. Of course they would fight.

But they, too, knew the siege had worked. Lee planned to leave Petersburg, march south, join with a Rebel army there, and continue the war. He thought he had a few weeks to get ready.

That was before George Pickett fought a battle at Five Forks. It was a place where roads come together near a railroad line—a very strategic spot. Lee told Pickett he had to hold Five Forks. Food and supplies for Petersburg and Richmond came down those roads and went through Five Forks. Pickett, the same man who led the famous charge at Gettysburg, did what he was told. He won the battle. But when it was over he went off to a shad bake a few miles away. (Shad is a delicious fish and Pickett was hungry.) While he was gone the Yankees decided to fight back. They counterattacked. This time they won. They captured the big prize: Five Forks. It was April Fool's Day, and when some Union soldiers were told of the victory, they couldn't believe the news; they thought they were being fooled.

Now there was no hope of the Confederates' surviving the siege. The next day, Robert E. Lee told Jefferson Davis to leave Richmond; then Lee and his army left Petersburg. Confederate troops decided to set fire to supplies in Richmond. They didn't want to leave anything for the Yankees. The flames soon spread. Food that might have fed hungry Richmonders was destroyed. People's homes were lost in the flames. Documents and records that went back to the founding of Jamestown were among the things burned.

When news of the fall of Richmond reached Washington, the citizens went wild with excitement. Cannons boomed, banners flew, and people hugged and kissed and cried with happiness. After four years of worry and gloom, finally the Confederate capital was captured.

On April 4, a tall man wearing a tall black hat stepped off a small military boat at the foot of one of Richmond's seven hills. He was escorted by 10 sailors. He had been told not to go to Richmond. It was too dangerous. He went anyway.

The men climbed two miles: past the governor's mansion, past Thomas Jefferson's capitol, to the White House of the Confederacy. Once, when they stopped to rest, an old black man with tears running

As Richmond burned, citizens bundled what they could into carriages and scrambled behind the troops across the James River to safety. "Disorder, pillage, shouts, mad revelry of confusion," said a reporter on the scene.

What has occurred in this case must ever recur in similar cases. Human nature will not change. In any future great national trial, compared with the men of this, we shall have as weak and as strong, as silly and as wise, as bad and as good. Let us therefore study the incidents of this, as philosophy to learn wisdom from, and none of them as wrongs to be revenged.

—ABRAHAM LINCOLN

> **Thank God I have lived to see this. It seems to me that I have been dreaming a horrid nightmare for four years, and now the nightmare is gone. I want to see Richmond.** —ABRAHAM LINCOLN, APRIL 3, 1865

Richmond, a reporter wrote, had "no sound of life, but the silence of the tomb....We are under the shadow of ruins."

> **I see the President almost every day....Mr. Lincoln...generally rides a good-sized, easy-going gray horse, is dressed in plain black, somewhat rusty and dusty, wears a black stiff hat and looks about as ordinary in attire, etc. as the commonest man....We have got so that we exchange bows and very cordial ones.**
> —WALT WHITMAN

down his cheeks came up to the tall man, took off his hat, bowed, and said, "May the good Lord bless you, President Lincoln." The president took off his hat and silently bowed back. It was a moment of affection between two men who understood each other.

All along the way blacks came to touch Lincoln and cheer him and sing to him. The white population stayed inside, watching through shuttered windows. Later, one of the president's sailor guards wrote of "thousands of watchers without a sound, either of welcome or hatred." The sailor said it was oppressive—eerie. But most of Richmond's whites were terrified. They had fought this man for four terrible years. Would he treat them as traitors? Lincoln entered the three-story brick mansion that had been the White House of the Confederacy. It was a fashionable house with columns at the entrance, gas lamps on the walls, and the latest in wallpaper. He sat in Jefferson Davis's chair, grinned, and said that he now felt like the president of all the United States.

Then he decided to try and find an old friend: General George Pickett. Lincoln had known George when he was a teenager. Pickett's uncle had been Lincoln's law partner. But, when he knocked at the

Pickett door, it was Mrs. Pickett who came to meet him. The general was not at home. Later, she described that meeting:

I had seen the carriage and the guard and retinue, but did not know who the visitors were. When I heard the caller ask for George Pickett's wife, I came forward with my baby in my arms.

"I am George Pickett's wife," I said.

"And I am Abraham Lincoln."

"The President?"

"No; Abraham Lincoln, George's old friend."

Seeing baby's outstretched arms, Mr. Lincoln took him, and little George opened wide his mouth and gave his father's friend a dewy baby kiss....As I took my baby back again, Mr. Lincoln said [to the baby] in that deep and sympathetic voice which was one of his greatest powers over the hearts of men:

"Tell your father, the rascal, that I forgive him for the sake of your mother's smile and your bright eyes."

Later Mrs. Pickett added:

I had sometimes wondered at the General's reverential way of speaking of President Lincoln, but as I looked up at his honest, earnest face, and felt the warm clasp of his great strong hand, I marvelled no more that all who knew him should love him.

But the war wasn't won yet. Lincoln knew that. Grant knew that, too. He knew he would have to capture or defeat Lee's army before it could end. That tired, shoeless army was now racing west. The soldiers were hungry, really hungry; some of them had gone days without any food at all. Their horses were half-starved, too. Luck wasn't with them. When they reached a railroad line, where food was expected, they found trainloads of ammunition. You can't eat bullets. They marched on.

Grant sent part of his army ahead; he was laying a trap. He wanted to surround Lee's army so there would be no way out. And finally he did that, at a place called Appomattox Court House.

Lincoln knew that in addition to their Kentucky birth, and their presidencies, he and Davis also had in common personal grief. The previous April, in 1864, Jefferson Davis's son, Joseph, had fallen from a high porch and been killed.

As Lincoln walked through the streets of Richmond he was surrounded by the city's former slaves. One old woman said, "I know I am free, for I have seen Father Abraham."

29 Mr. McLean's Parlor

"I have probably to be Gen. Grant's prisoner," said Lee, "and thought I must make my best appearance."

Wilmer McLean didn't like to be hassled. So when he retired from business he bought a comfortable farm with pleasant fields, woods, and a stream. He planned to live there quietly with his family. McLean's farm was in Virginia, but not far from Washington. It was near an important railroad junction. The stream that crossed the farm was called *Bull Run*.

Do you think you know what happened on his farm? Well, you don't know all of it. Let's go back in time—to 1861. In April, you remember, the war began when Confederates in Charleston, South Carolina, fired their cannons at Fort Sumter. We are now in July of 1861. As yet there have been no big battles.

The two armies—North and South—are gathered near Manassas Junction. The Confederates are using Wilmer McLean's farm as a meeting place. One day some Southern officers are about to have lunch with the McLeans when the Union artillery zooms a cannonball at the house. It goes straight through the roof and lands in a kettle of stew. The shell explodes, so does the kettle, and stew is spattered all over the room!

That is just the beginning of farmer McLean's troubles. Because of its railroad lines, Manassas is a strategic spot. Neither army wants the other to control it. After the battle

Wilmer McLean, destined to be "in at the beginning" (left, his Manassas house, where a cannonball fell through the roof) and...

of Bull Run, soldiers stay around, and, a year later, another battle is fought there. Wilmer McLean has had enough. He decides to move someplace very quiet. He wants to be as far from the war as possible. So he moves to a tiny, out-of-the-way village called Appomattox Court House.

Maybe Wilmer McLean had magnets in his blood: he seemed to attract historic occasions. In 1865 the two armies—North and South—found themselves at Appomattox Court House. A Confederate officer was looking for a place to have an important meeting. Wilmer McLean showed him an empty building. It wouldn't do. So McLean took him to his comfortable red-brick house. That turned out to be just fine. It was in Wilmer McLean's front parlor that Robert E. Lee officially surrendered to Ulysses S. Grant. In later years McLean is supposed to have said, "The war began in my dining room and ended in my parlor."

On April 9, 1865, Robert E. Lee—proud, erect, and wearing his handsomest uniform—walked into Wilmer McLean's parlor. Strapped to his side was a gorgeous, shining sword with a handle shaped like a lion's head. The sword was decorated with carvings and held in a fine leather scabbard.

General Grant and his aides couldn't help looking at the beautiful sword. They all knew that, according to the rules of war, the defeated general must give his sword to the winner.

General Lee knew that too. But he was not the kind of person who would bring an old sword to give away. He had brought his most precious sword. He had worn his best uniform. He held his head high. He knew he had fought as hard as he could. He had lost the war—fair and square—but he had not lost pride in himself and his men. Robert E. Lee's dignity and courage would be an example to his men when they returned to their homes. They didn't need to apologize for themselves, they

..."in at the death" of the Civil War, at his new home (above) in Appomattox Court House.

Making Amends

When the Constitution was written, back in 1787, it had a terrible flaw. (What was that terrible flaw?) After the Civil War three amendments to the Constitution—the 13th, 14th, and 15th—were passed. They corrected the flaw and did something else, too. The amendments added more of the spirit of the Declaration of Independence to the Constitution.

The 13th Amendment, adopted in 1865, prohibited slavery.

The 14th Amendment, adopted in 1868, gave equal protection of the law to ALL Americans.

The 15th Amendment, passed in 1870, said that all citizens have the right to vote. Are women citizens? This amendment didn't say yes and it didn't say no, which was too bad. It would take another amendment to make that clear.

Black men voting after the war. The vote gave blacks real power. Then (until well into the 20th century) they lost the vote and their power.

Lee signs the surrender terms in the McLeans' parlor. "What General Lee's feelings were, I do not know," Grant wrote later. "As he was a man of much dignity, with an impassible face...his feelings...were entirely concealed from my observation."

had fought as well as men can fight.

General Grant understood that. Later he described his feelings on that day.

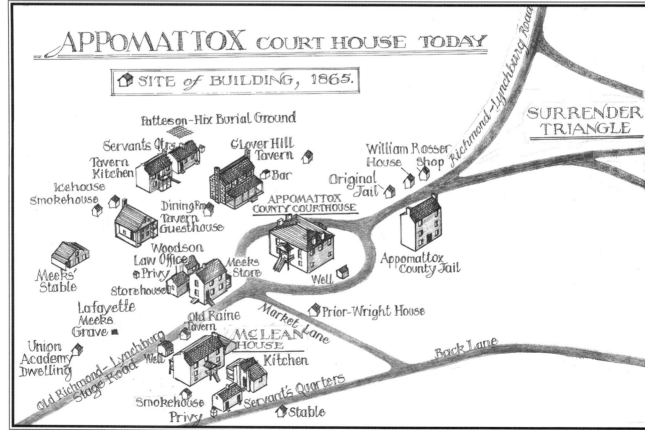

APPOMATTOX COURT HOUSE TODAY

SITE of BUILDING, 1865.

Patteson-Hix Burial Ground

Servants Qtrs.

Clover Hill Tavern

Bar

William Rosser House Shop

Richmond-Lynchburg Road

SURRENDER TRIANGLE

Tavern Kitchen

Original Jail

APPOMATTOX COUNTY COURTHOUSE

Icehouse Smokehouse

Dining Rm Tavern Guesthouse

Appomattox County Jail

Woodson Law Office

Privy

Meeks Store

Well

Meeks' Stable

Storehouse

Prior-Wright House

Lafayette Meeks Grave

Old Raine Tavern

Market Lane

Back Lane

Union Academy Dwelling

Old Richmond-Lynchburg Stage Road

Well

McLEAN HOUSE

Kitchen

Smokehouse Privy

Servant's Quarters

Stable

I felt...sad at the downfall of a foe who had fought so long and so valiantly, and had suffered so much for a cause, though that cause was, I believe, one of the worst for which a people ever fought.

But what should Grant do with Lee's sword? Keep it as a treasure to give to his children and grandchildren? Turn it over to the country to put in a museum?

Ulysses S. Grant didn't do either of those things. He wrote out the official surrender terms. They were kinder than anyone had expected. The Southern soldiers could go home, and—as long as they gave their promise not to fight against the country again—they would not be prosecuted for treason. They must surrender their guns, but could take their horses. General Grant inserted a phrase in the document: the surrender did not include "the side arms of the officers." No one in the room said anything about it, but they all knew: Lee's sword would stay strapped to his side.

But there is something more important than a sword to remember about that ceremony in McLean's parlor. There were important words said there, and General Grant didn't say them; nor did General Lee. They came after the signing of the papers, when there were handshakes all around. General Lee was introduced to General Grant's staff. One of Grant's aides was copperskinned Lieutenant Colonel Ely Parker. Robert E. Lee looked at him for a moment and said, "I am glad to see one real American here." Parker—a Seneca Iroquois—replied firmly, "We are all Americans."

Robert E. Lee—brave and heroic as he was—still didn't

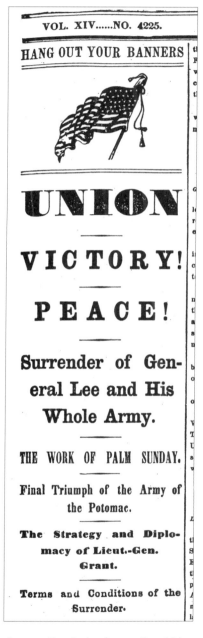

Appomattox today is a national historic site. At Surrender Triangle the Confederates laid down their arms.

In a painting made long after the war was over, men of Lee's Army of Northern Virginia cry as they furl the Stars and Bars for the last time.

We Are All Americans

Ely Parker's real name was Donehogawa, and he was Keeper of the Western Door of the Long House of the Iroquois. As a boy, Donehogawa decided that he wanted to be successful in the world of white men and women. So he chose a white man's name for himself, and studied law. But when it came time to be admitted to the bar and become a lawyer, he was told that only white males were acceptable. Parker/Donehogawa was not the kind of person who gave up. He decided to try another profession. He went to college, became an engineer, and helped supervise the building of the Erie Canal. His work took him to Galena, Illinois, where he became friends with a clerk in a harness shop who was without prejudice. The clerk was named Ulysses S. Grant. When Grant became General Grant and needed an engineer he turned to his friend Ely Parker.

After the war, Parker/Donehogawa was a brigadier general; he became commissioner of Indian affairs for the government. He tried his best to help his people. United States citizens were pushing west and taking the Indian lands. No one—not even Ely Parker—could stop them. Parker resigned as Indian affairs commissioner and became wealthy as a Wall Street investor.

seem to understand why so many men and women had been willing to fight and sacrifice and die in this terrible war. *We are all Americans.* It was in those words.

Ely Parker knew that we aren't all the same. Our skins are different colors, our religions are different, our abilities are different, our backgrounds may be different. So what is it that makes us the same? What is it that makes us *all Americans*?

An idea. We share an idea. That's what makes us alike. Other nations didn't begin with ideas; most began with barons and kings.

We started with a declaration that said *all men are created equal.* That new and powerful idea excited people all over the world. But our Constitution had not guaranteed that equality. This Civil War—terrible as it was—caused the Constitution to be changed for the better. Three constitutional amendments—the 13th, 14th, and 15th—would soon be passed. They would make sure that *we are all Americans.* They would give the nation *a new birth of freedom.*

30 A Play at Ford's Theatre

"When Grant gets possession of a place," said Lincoln, "he holds on to it as if he had inherited it."

Abraham Lincoln was in a good mood. The vacation on the *River Queen*, and the exciting events of the past week, had given him back his energy and optimism. It was April 14, 1865—and he was meeting with his cabinet. General Grant, now a great war hero, was a guest at the meeting. The two men enjoyed being together, and, since the war was just about over, they could joke a bit and relax.

The president told Grant and the cabinet members of a dream he had the previous night. He was on a boat heading for a distant, misty shore. It was not such an unusual dream, but what was unusual was that he had had that same dream before. Each time he dreamt of the boat and the misty shore he learned big war news. So now he was sure the nation would hear something important before the day was over. Perhaps, he said, General Sherman had captured the last remaining Confederate army.

The cabinet members were in a good mood, too. They laughed about the dream and went on to serious matters. It was Reconstruction they talked about. How would the South be brought back into the Union? That was the important problem facing the nation. What should be done to help the newly freed men and women become useful citizens? They needed schooling, they needed land to farm, they needed jobs. How were the defeated white Southern leaders to be treated? After four years of terrible war many Northerners were in no mood to forgive them. Some wanted the

Reconstruction was the name Lincoln used for the process of bringing North and South together into a united nation. What should be done with the officers of the Confederacy? Were they traitors? Traitors were usually hanged. What about the newly freed men and women? There was much to be decided.

In *Moving Day for the Confederacy,* the cabinet flees, its treasury and arms chests empty.

Rock of Ages

On February 1, 1865, President Lincoln signed a congressional resolution (adopted by both houses of Congress) that proposed a 13th amendment outlawing slavery. That same day, Dr. John Swett Rock (who was both a lawyer and a physician) was admitted to practice before the U.S. Supreme Court. Moments after that, Dr. Rock was "received upon the floor of the House of Representatives while it was in session," the first black person to be so honored.

The Congress, and most observers, were aware of the momentousness of the event. This was the Supreme Court that just a few years earlier (when led by Chief Justice Roger B. Taney) had declared that Dred Scott—and all blacks—had no rights as American citizens. But Salmon P. Chase was now chief justice. Rock described him as "a great and good man," and said, "with him I think my color will not be a bar to my admission [to the Supreme Court]." Senator Charles Sumner said, "the admission of a colored lawyer to the bar of the Supreme Court would make it difficult for any restriction on account of race to be maintained anywhere." The court was repudiating Taney's decision when it honored Dr. Rock. Everyone understood that. (*Repudiating* means rejecting or casting off a previous decision.)

But, when he left the capital to return to Boston, John Rock was just another black to the train-ticket taker—and he didn't have a pass to leave the city. (Blacks needed identification passes; whites did not.) He had to get a pass before he could go home. Senator Sumner wrote that Rock's admission to the court "helped the way for admission of his race to the rights of citizenship, and especially the right to vote." But many other "great and good" people would be needed to carry on the fight for genuine equality of rights.

Southern leaders hanged. There was much talk but no decisions. Lincoln listened. He would announce his plans soon, he said.

The war had changed the president. He was not the same man he had been four years earlier. Everyone could see that. Lincoln had suffered terribly: his son had died, he had seen a generation of young Americans die. He had a spiritual quality now—some people called it a kind of saintliness. Whatever it was, he seemed to have found peace within himself.

He had always believed that blacks should be treated like whites. It was the only fair thing. And he was a fair-minded man. Before the war he had had suspicions that the races were different. That maybe blacks did not have the same needs, desires, and abilities as whites. That was before he got to know some black leaders. Before he listened to black soldiers. Before a black woman became his wife's friend, and his. Now he knew differently. He planned to use that knowledge to make the country wiser, and better.

He knew that while the cabinet was meeting, an American flag was being raised at Fort Sumter in the harbor at Charleston, South Carolina. The flag was the very one that had been lowered exactly five Aprils earlier. It was bullet-torn. Shots fired at that flag at Fort Sumter had begun the war. Now there would be cheers for the flag and there would be words of peace spoken. The nation's best-known abolitionist, William Lloyd Garrison, was in Charleston. So was the Reverend Henry Ward Beecher, who was Harriet Beecher Stowe's brother and the country's most famous preacher. Northerners were going south again—and not as soldiers. Some had already opened schools for the newly freed men, women, and children. Doctors and nurses were helping the sick and wounded.

In Washington, at the cabinet session, someone handed General Grant a note. It was from his wife. The note said they were to go to Philadelphia to be with their sons. President Lincoln was disappointed. He had looked forward to spending the evening with the general. The Grants had been invited, along with the Lincolns, to a play at Ford's Theatre, the popular Washington playhouse located between the White House and the Capitol. Lincoln enjoyed going to the theater, although this was said to be a silly play. When he heard Grant was not going, the president told an aide he didn't want to go either. But he knew Mrs. Lincoln was looking forward to an evening out, and he didn't want to disappoint her.

So they went, with some other guests, and sat in the flag-covered president's box. The box was a small, separate balcony that hung over the stage. It was a good place to see the play, and, even though the audience couldn't quite see Lincoln behind the curtains and flags, they saw him enter and they stood while the orchestra played "Hail to the Chief." The play was funny, and Abraham Lincoln enjoyed laughing, so he must have relaxed in the comfortable chair the managers had put in the box especially for him.

Then something happened. It was as if the play shifted to the president's box. No one could believe what they saw and heard. Perhaps it was all part of the act. That was what some of the people in the audience thought. There was a sound like a small thunder boom, and some smoke. Then a wild-acting man climbed up out of the president's box, leaped onto the stage, said something in Latin, and was gone. A woman screamed and a voice cried out, *The*

Top, Ford's Theatre; above, John Wilkes Booth; left, a poster advertising Lincoln's presence at the play.

What the assassin is believed to have said was *Sic semper tyrannis*—"thus ever to tyrants"—which means "This is the way tyrants are treated." It is Virginia's state motto.

The giant sufferer lay extended diagonally across the bed, which was not long enough for him....Robert, his son, stood at the head of the bed. He bore himself well, but on two occasions gave way...and sobbed aloud...leaning on the shoulder of Senator Sumner.

—GIDEON WELLES

The soldiers are wild with rage to think that this great and good man who did so much for our land should be stricken down in the hour of victory.

—ELISHA RHODES

Mother prepared breakfast—and other meals—as usual; but not a mouthful was eaten all day by either of us. We each drank half a cup of coffee; that was all. Little was said. We got every newspaper, morning and evening...and passed them silently to each other.

—WALT WHITMAN, APRIL 16, 1865

president has been shot. And everyone knew that this was no act. It was real.

Abraham Lincoln died the next day in the place he was carried to—a small house across the street from the theater. It was April 15, 1865.

That very same day Andrew Johnson was

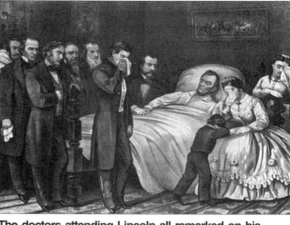

The doctors attending Lincoln all remarked on his splendid, muscular physique, the body of a much younger man than his deeply lined face suggested.

sworn in as president. A month and a half earlier, on the day of Lincoln's second inauguration, when Johnson was named vice president, he had embarrassed everyone by being drunk.

The Whole World Bowed in Grief

A few hours after her husband's death, Mary Todd Lincoln asked Elizabeth Keckley to come and be with her. Keckley, born a slave, was a successful Washington seamstress, a friend of Mary Lincoln, and an almost daily visitor to the White House. Here, Keckley writes of that time:

When [Mrs. Lincoln] became a little quiet, I asked and received permission to go into the Guests' Room, where the body of the President lay in state. When I crossed the threshold of the room, I could not help recalling the day on which I had seen little Willie lying in his coffin where the body of his father now lay. I remembered how the President had wept over the pale beautiful face of his gifted boy, and now the President himself was dead. The last time I saw him he spoke kindly to me, but alas! the lips would never move again. The light had faded from his eyes, and when the light went out the soul went with it!

What a noble soul was his— noble in all the noble attributes of God. Never did I enter the solemn chamber of death with such palpi- tating heart and trembling footsteps as I entered it that day. No common mortal had died. The Moses of my people had fallen in his hour of triumph....When I entered the room, the members of the Cabinet and many distinguished people were grouped around the body of their fallen chief. They made room for me, and, approaching the body, I lifted the white cloth from the white face of the man that I had worshipped as an idol....There lurked the sweetness and gentleness of childhood, and the stately grandeur of godlike intellect. I gazed long at the face, and turned away with tears in my eyes and a choking sensation in my throat. Ah! never was a man so widely mourned before. The whole world bowed their heads in grief when Abraham Lincoln died.

31 After Words

War Department, Washington, April 20, 1865,

$100,000 REWARD!

THE MURDERER

Of our late beloved President, Abraham Lincoln,

IS STILL AT LARGE.

$50,000 REWARD

$25,000 REWARD

$25,000 REWARD

EDWIN M. STANTON, Secretary of War.

The murder was actually a conspiracy among several people, including one woman.

Oh, murder most foul! Oh, woe! Oh, tragedy! Never was a murder more terrible for a nation. In ancient Rome, Caesar was cruelly killed and the flow of history changed. This was worse. A 26-year-old actor had shot a great and good president just when he was most needed.

John Wilkes Booth—the assassin—had not fought in the war and was plagued with feelings of cowardice. He wanted to feel important. He called Lincoln a tyrant. He was sure people in the South would cheer his act and think him a hero. They didn't. Confederate General George Pickett said, "The South has lost her best friend and protector in this her direst hour of need."

John Wilkes Booth was hunted and trapped in a Virginia tobacco barn. The barn was set on fire. When Booth wouldn't come out, he was shot.

Southerners hadn't wanted Lincoln assassinated. He was their president too. In Norfolk, Virginia, on the day of Lincoln's funeral, a long procession of people marched through the streets while a military band played sad music. Many of the marchers were former Confederate soldiers.

Perhaps they understood that though the war was over, peace was yet to be won. Forces of greed and cruelty and bewilderment were readying themselves for new kinds of battle. The tall president in his tall hat had been strong enough to withstand them. He had said, "Let us therefore study the incidents of this [war], as philosophy to learn wisdom from, and none of them as wrongs to be revenged." He had said *with malice toward none*, and meant it.

The new president, Andrew Johnson, was an ordinary man. He didn't know how to fight the demons of hate. He wasn't strong enough to subdue the monsters of revenge. Oh, misguided murderer. You, John Wilkes

> [The war] in a sense was begun by one madman, John Brown, and ended by another, John Wilkes Booth.
>
> —SHELBY FOOTE

> If men were equal in America, all those former Poles and English and Czech and blacks, then they were equal everywhere, and there was really no such thing as a foreigner; there were only free men and slaves. And so it was not even patriotism but a new faith. The Frenchman may fight for France, but the American fights for mankind, for freedom; for the people, not the land.
>
> —MICHAEL SHAARA, *THE KILLER ANGELS*

Vice President Johnson was a Tennessee senator who refused to join the Confederacy when his state seceded.

Booth, were the tyrant and the villain! The people you hurt most were the people of the South.

Without the good president, Reconstruction didn't turn out the way Lincoln intended. He wanted the nation's wounds bound carefully. He wanted healing to take place. He wanted North and South to be one united nation. He wanted those who had been slaves to be treated like full citizens. He knew none of that would be easy, but he loved and understood both North and South—most of the political leaders who came after him didn't. They took sides. They messed things up.

Still, the war had accomplished much. The United States had withstood the terrible fire of civil war. Lincoln had seen that war as a test of free, democratic government. Could people rule themselves? Could they put down a rebellion from within the country? Well, they had. The Union had held. Its citizens had been willing to die to save it. They had shown the strength of the American vision.

What seemed to be remarkable was that when the war was over, it really was over. There was no more fighting. In other

Lincoln's funeral train crossed the country from Washington, D.C., to Springfield, Illinois. In Chicago the hearse, drawn by horses, carried the coffin in procession under this arch draped in black crepe to the courthouse, where the president lay in state overnight. At right, the hanging of the four remaining conspirators (Booth had already been shot).

lands, rebel forces often took to the hills, became guerrilla fighters, and carried on for years. Some Confederate soldiers had said they would go into the hills and fight on, "like rabbits and partridges." But General Lee told them to go home, plant crops, and be good citizens—and that was what they did.

Something else was remarkable. Lincoln had said, "This nation under God shall have a new birth of freedom." And it happened. The amendments—the 13th, 14th, and 15th—made the Constitution stronger and better. They made right those things that had not been right in the Constitution of 1789. Slavery was no more. The new amendments said the Constitution and the Bill of Rights were meant to protect all Americans.

No, the evils of racism weren't finished; they would still cause much pain and evil. But laws were in place that said clearly: unfairness was wrong. Black men, by law, were full citizens. (Women were now demanding equal rights. If black men could be voting citizens, why not women—black and white? Blacks had known they were as smart and capable as whites. Women knew they were as smart and capable as men.)

It would be more than a century before the wounds of war were fully healed. Americans would slowly come to understand that the Declaration of Independence, the Constitution, the Bill of Rights, and the 13th, 14th, and 15th amendments were all parts of a great, unfinished quest. It was a quest for a just society. Nowhere had there ever been such a society. Perhaps it would never be achieved. But Americans are believers. More than 600,000 people—mostly young, idealistic men—had died because they believed in freedom. Almost half a million would live the rest of their lives with a missing arm or leg or a painful wound. Would their suffering have meaning? Yes, it would. Because of them the quest would continue, although the path would be far from straight.

When you are old enough to vote, remember the awful consequences when people follow poor leaders. Some Southern whites had tried to prevent the Civil War—but few people listened to them. After the war, others tried to make the South fair and just for all people; they were ridiculed and called "scalawags." Real leaders push people forward: toward better ways. But there didn't seem to be any real leaders in the South. Up North, things weren't much better.

Many Northern politicians wanted to punish the South. There were few with generous hearts. What was missing was a great national leader who could end the bitterness: someone who spoke eloquently and would remind North and South that it was a Southerner who had written that *all men are created equal*. Oh, tragedy! The leader who might have done that had been killed by a 26-year-old actor.

Jefferson Davis was captured and imprisoned at Fortress Monroe in Hampton, Virginia. He was said to have dressed in his wife's clothes to escape detection, but this was only rumor. When he was ordered to be put in chains he protested, saying, "Those are orders for a slave, and no man with a soul in him would obey such orders." Years later, talking to a reporter, Davis said, "Tell the world that I only loved America."

Of all the men I ever met, he [Lincoln] seemed to possess more elements of greatness, combined with goodness, than any other.

—General W. T. Sherman

Songs of the Civil War

Today, if you want to hear music, you can turn on your headset. So it is hard for most of us to understand the importance of singing in the Civil War. But the soldiers sang everywhere. They sang when they marched, they sang around the campfires, and they sang on the battlefield—to cheer themselves when things were going well, or to rally themselves when things were going badly. Songs helped them survive in the hideous prisons. And it helped them to know that at home their families were singing the same songs.

How do you think listening to music on a headset compares to group singing? Try singing some of these songs with your friends—then close your eyes and imagine yourself sitting around a campfire, or marching, or fighting.

Everyone in the North sang "John Brown's Body." It is a great marching song, set to the same tune as the "Battle Hymn of the Republic." Besides, John Brown was an abolitionist hero. What hardly anyone knew was that the song was not written about the John Brown who led the rebellion at Harper's Ferry. It was actually a spoof—a sort of joke or takeoff—written by the men of the 12th Massachusetts Regiment about their own Sergeant John Brown!

Marching Through Georgia
WORDS AND MUSIC BY HENRY CLAY WORK

Bring the good old bugle, boys!
We'll sing another song;
Sing it with a spirit
That will start the world along;
Sing it as we used to sing it,
Fifty thousand strong,
While we were marching through Georgia.

CHORUS:
"Hurrah, hurrah! We bring the Jubilee!
Hurrah, hurrah! The flag that makes you free!"
So we sang the chorus
From Atlanta to the sea,
While we were marching through Georgia.

Yes, and there were Union men
Who wept with joyful tears
When they saw the honor'd flag
They had not seen for years;
Hardly could they be restrained
From breaking forth in cheers,
While we were marching through Georgia!
CHORUS

"Sherman's dashing Yankee boys
Will never reach the coast!"
So the saucy Rebels said,
And 'twas a handsome boast,
Had they not forgot, alas,
To reckon with the host,
While we were marching through Georgia!
CHORUS

So we made a thoroughfare
For Freedom and her train,
Sixty miles in latitude,
Three hundred to the main;
Treason fled before us,
For resistance was in vain,
While we were marching through Georgia!
CHORUS

Northern soldiers marched home singing this song, which was written at the end of the war. Later on the tune was used as a marching song by the Japanese army when it invaded Manchuria in 1933, the British army when it occupied India, and again by the U.S. Army in the North African desert during World War II.

Dixie's Land

WORDS AND MUSIC BY DANIEL DECATUR EMMETT

I wish I was in the land of cotton,
Old times there are not forgotten;
Look away, look away, look away, Dixie Land!
In Dixie Land, where I was born in,
Early on one frosty mornin';
Look away, look away, look away, Dixie Land!

CHORUS:
Then I wish I was in Dixie, hooray, hooray!
In Dixie Land I'll take my stand,
To live and die in Dixie;
Away, away, away down south in Dixie;
Away, away, away down south in Dixie.

Old Missus marry Will the weaver,
Willium was a gay deceiver;
Look away, etc.
But when he put his arm around 'er,
He smiled as fierce as a forty-pounder;
Look away, etc.
CHORUS

His face was sharp as a butcher's cleaver,
But that did not seem to grieve 'er;
Look away, etc.
Old Missus acted the foolish part,
And died for a man that broke her heart;
Look away, etc.
CHORUS

Now, here's a health to the next Old Missus,
And all the gals that want to kiss us;
Look away, etc.
But if you want to drive 'way sorrow,
Come and hear this song tomorrow;
Look away, etc.
CHORUS

There's buckwheat cakes and Injun batter
Makes you fat or a little fatter;
Look away, etc.
Then hoe it down and scratch your gravel,
To Dixie's Land I'm bound to travel;
Look away, etc.
CHORUS

Goober Peas

WORDS AND MUSIC BY ARMAND E. BLACKMAR

Sitting by the roadside on a summer day,
Chatting with my messmates, passing time away,
Lying in the shadow underneath the trees,
Goodness, how delicious, eating goober peas!

CHORUS:
Peas! Peas! Peas! Peas! Eating goober peas!
Goodness, how delicious, eating goober peas!

When a horseman passes, the soldiers have a rule:
To cry out at their loudest, "Mister, here's your mule!"
But another pleasure, enchantinger than these,
Is wearing out your grinders eating goober peas!
CHORUS

Just before the battle the gen'ral hears a row;
He says, "The Yanks are coming, I hear their rifles
 now."
He turns around in wonder, and what d'you think he
 sees?
The Georgia militia, eating goober peas!
CHORUS

I think my song has lasted almost long enough,
The subject's interesting but the rhymes are mighty
 rough;
I wish this war was over, when free from rags and
 fleas,
We'd kiss our wives and sweethearts, and gobble
 goober peas!
CHORUS

The Battle Hymn of the Republic

WORDS BY JULIA WARD HOWE

Mine eyes have seen the glory of the coming of the Lord;
He is trampling out the vintage where the grapes of
*   wrath are stored;*
He hath loosed the fateful lightning of his terrible swift
*   sword,*
His truth is marching on.

CHORUS:
Glory, glory hallelujah!
Glory, glory hallelujah!
Glory, glory hallelujah!
His truth is marching on.

I have seen Him in the watch fires of a hundred circling
*   camps;*
They have builded Him an altar in the ev'ning dews and
*   damps;*
I can read his righteous sentence by the dim and flaring
*   lamps,*
His day is marching on.
CHORUS

I have read a fiery gospel writ in burnish'd rows of steel;
"As ye deal with My contemners, so with you My grace
*   shall deal";*
Let the Hero born of woman crush the serpent with his
*   heel,*
Since God is marching on.
CHORUS

He has sounded forth the trumpet that shall never call
*   retreat;*
He is sifting out the hearts of men before His judgment
*   seat.*
Oh, be swift, my soul, to answer Him! Be jubilant, my
*   feet!*
Our God is marching on.
CHORUS

In the beauty of the lilies Christ was born across the sea,
With a glory in His bosom that transfigures you and me;
As He died to make men holy let us die to make men free,
While God is marching on.
CHORUS

John Brown's Body

John Brown's body lies a'molderin' in the grave,
John Brown's body lies a'molderin' in the grave,
John Brown's body lies a'molderin' in the grave,
His soul is marching on.

CHORUS:
Glory, glory hallelujah!
Glory, glory hallelujah!
Glory, glory hallelujah!
His soul is marching on.

The stars of heaven are looking kindly down
[THREE TIMES]
On the grave of old John Brown.
CHORUS

He's gone to be a soldier in the army of the Lord
[THREE TIMES]
His soul is marching on.
CHORUS

John Brown's knapsack is strapped upon his back
[THREE TIMES]
His soul is marching on.
CHORUS

His pet lambs will meet him on the way
[THREE TIMES]
And they'll go marching on.
CHORUS

They will hang Jeff Davis to a sour apple tree
[THREE TIMES]
As they go marching on.
CHORUS

When Johnny Comes Marching Home

WORDS AND MUSIC BY PATRICK S. GILMORE

When Johnny comes marching home again, hurrah, hurrah,
We'll give him a hearty welcome then, hurrah, hurrah;
The men will cheer, the boys will shout,
The ladies, they will all turn out,
And we'll all feel gay when Johnny comes marching home.

Chronology of Events

1820: Under the Missouri Compromise, slavery is banned north of 36° 30' except in Missouri

1850: Under the Compromise of 1850, some western territories may choose whether to permit slavery, and the Fugitive Slave law forbids helping runaways

1852: Harriet Beecher Stowe publishes *Uncle Tom's Cabin*

1854: Kansas–Nebraska Act splits western settlers

1857: the *Dred Scott* decision

1859: John Brown attacks Harper's Ferry and is executed

Nov. 1860: Abraham Lincoln elected 16th president

Dec. 1860: South Carolina secedes from the Union

Jan.–Feb. 1861: Mississippi, Florida, Alabama, Georgia, Louisiana, and Texas secede

Feb. 1861: provisional government of Confederate States of America established; Jefferson Davis is elected president of the Confederacy

Apr. 1861: Fort Sumter surrenders to Confederacy

Apr. 1861: Lincoln declares insurrection; Federal blockade of Southern ports begins

Apr.–May 1861: Virginia, Arkansas, Tennessee, and North Carolina secede

May 1861: Richmond, Virginia, becomes Confederate capital

July 1861: U.S. troops defeated at Bull Run

Aug. 1861: Confederates defeat Union in the West at battle of Wilson's Creek, Missouri

Nov. 1861: Gen. George McClellan replaces Gen. Winfield Scott as U.S. Army commander in chief

Feb. 1862: Union Gen. U. S. Grant captures Ft. Henry and Ft. Donelson in Tennessee

Mar. 1862: battle of *Monitor* and *Merrimack*

Mar. 1862: Gen. Robert E. Lee becomes effective commander in chief of Confederate army

Mar.–July 1862: McClellan attempts to reach Richmond in Peninsular campaign

Mar.–May 1862: Gen. T. J. "Stonewall" Jackson drives Union forces from Shenandoah Valley

Apr. 1862: U. S. Grant escapes defeat at Shiloh

Apr. 1862: conscription begins in Confederacy

Apr. 1862: New Orleans surrenders to Union navy under Adm. David Farragut

June–July 1862: McClellan fails to take Richmond in Seven Days' battles

Aug. 1862: Union again defeated at Bull Run

Sept. 1862: terrible casualties at battle of Antietam in Maryland; Lee retreats to Virginia

Sept. 1862: Lincoln issues Emancipation Proclamation; slavery illegal in Rebel states

Mar. 1863: U.S. begins conscription

May 1863: Confederates win at Chancellorsville but Stonewall Jackson dies, shot by friendly fire

May 1863: Grant besieges Vicksburg, Mississippi

July 1863: Confederates defeated at battle of Gettysburg, Pennsylvania; terrible losses sustained on both sides in a turning point of the war

July 1863: Vicksburg surrenders to Grant; Union now controls Mississippi River

Sept.–Nov. 1863: U.S. armies under Grant gain control of much of Western theater

Mar. 1864: Grant named U.S. commander in chief

May 1864: Gen. W. T. Sherman begins march through Georgia to split the South

June 1864–Apr. 1865: Grant besieges Petersburg

July–Sept. 1864: Sherman besieges and captures Atlanta, Georgia

Sept.–Oct. 1864: Union General Philip Sheridan devastates Shenandoah Valley

Nov. 1864: Lincoln reelected as U.S. president

Nov.–Dec. 1864: Sherman's March to the Sea

Jan.–Mar. 1865: Sherman's drive through Carolinas

Mar. 1865: Lincoln's second inaugural

Apr. 1865: fall of Petersburg and Richmond, Virginia

Apr. 1865: Lee surrenders to Grant at Appomattox Court House, Virginia

Apr. 1865: Lincoln assassinated by John Wilkes Booth; Andrew Johnson named 17th president

May 1865: the last armed Confederate force surrenders in New Orleans

More Books to Read

Joy Hakim asked me to make suggestions for other books about the Civil War. All these books are good— and fun to read. —TAMARA GLENNY, EDITOR

Patricia Clapp, *The Tamarack Tree,* Lothrop, Lee & Shepard, 1986. English-born Rosemary Leigh is 13 when her mother dies, leaving Rosemary and her older brother Derek to find their own way to relatives in Vicksburg, Mississippi. Rosemary makes friends with a Southern belle and falls for a Yankee boy—but now Vicksburg is under siege and its citizens are starving. What will become of them?

Stephen Crane, *The Red Badge of Courage,* Puffin, 1986. Stephen Crane died aged 28, but not before producing (in 1895) the most famous book ever written about the Civil War. The story of Henry Fleming, a raw young recruit in the Union army, and the horrors he sees and endures, is as immediate, harrowing, and moving as if the author had actually been there. This great book was written for adults, but it is short and not hard to read.

Paul Fleischman, *The Borning Room,* Harper Collins, 1991. The "borning room" is an alcove off the kitchen where babies are born in the Ohio farmhouse that Georgina Lott's grandpa built. As the war rages, Georgina grows up. She goes to school, helps a runaway slave, falls in love, and gives birth herself in the borning room. This lovely book reminds us that even in wartime people live, play, work, and die as always.

Irene Hunt, *Across Five Aprils,* Follett, 1964. Jethro Creighton is only nine as the war begins, but he has to grow up fast when his brothers leave their Illinois homestead to fight—not all of them on the Union side. His father has a stroke after a mob burns their barn, and Jethro must quit school and farm the land. From letters, rumors, and news accounts the family lives through the war's five Aprils—and so does the reader of this wonderful story.

Harold Keith, *Rifles for Watie,* HarperCollins, 1957. When the author of this book was a young man he interviewed several Oklahoma veterans of the Civil War's western campaigns; years later he wrote this exciting, realistic story of Kansas farm boy Jefferson Davis Bussey, who despite his name joins the Union army. Sent to spy on Cherokee Indian troops fighting for the South under Chief Stand Watie, Jeff is caught and has to pretend to be a Confederate soldier.

Milton Meltzer, *Voices From the Civil War,* HarperCollins, 1989. A collection of original documents—from diaries, songs, letters, newspapers, speeches. Here are fuller versions of many of the words quoted in *A History of US.*

Jim Murphy, *The Boys' War,* Clarion, 1990. Using the words, stories, and photographs of real boys and young men who fought in the Civil War, the author of this excellent book gives us a true feeling of what it was like to be there. Also included is the most useful bibliography of books about the war that I have seen.

Carolyn Reeder, *Shades of Gray,* Macmillan, 1989. The fighting is over when this story begins—but not for 12-year-old Southerner Will Page, orphaned by the war. Will must leave his home to live with family in the Virginia Piedmont. Uncle Jed refused to fight for the Confederacy on principle—but Will endures a lot of anger and misery before he begins to understand that his uncle, too, had to be brave to make his choice.

Ann Rinaldi, *The Last Silk Dress,* Holiday House, 1988. The author found a letter from Rebel General Longstreet telling of a spy balloon that the ladies of Richmond made by giving away their silk dresses. Then she imagined the story of Susan Chilmark, the girl who might have collected those dresses—and of how she came to understand some of the war's complexity. This is a long, enjoyable, romantic book.

G. Clifton Wisler, *Red Cap,* Lodestar, 1991. Ransom Powell was a real boy from Maryland who joined the Union army as a drummer in 1862, aged 13. Two years later, he was captured and taken to the Confederate prison camp at Andersonville, Georgia. As camp drummer, Red Cap survived (barely) to tell the tale of Andersonville and to see 13,000 of his fellow inmates die of starvation, disease, and despair. This is a memorable, inspiring book about a dreadful part of the war.

Index

A

abolitionism, 12, 13, 23–26, 27, 32, 43, 47, 52, 53, 61, 75, 100, 103, 144
Adams, Charles Francis, Jr., 97, 132
Africa, 49, 50, 51
African Americans. *See* blacks
Africans, 12
agriculture, 11, 50, 52, 60, 62, 64, 73, 79, 81, 103. *See also* crops; farming
Alabama, 59, 77, 97
Alabama (ship), 94, 95
amendments to the Constitution. *See* constitutional amendments
American Indians. *See* Indians; Native Americans
American Red Cross, 99, 111
American Revolution. *See* Revolutionary War
ammunition, 55, 83, 84, 97, 112, 113, 120, 126, 137
Anaconda Plan, 65, 67. *See also* strategy
Anderson, Maj. Robert, 18
animals, 36, 61, 62, 64, 65, 70, 84, 90, 111, 115, 117, 120, 126
anti-slavery. *See* abolitionism
Antietam, battle of, 14, 98, 99, 109. *See also* Sharpsburg, Maryland
Appomattox Court House, Virginia, 137–142
Appomattox River, 125
Arkansas, 59, 64
Armistead, Capt. Lewis, 77, 116
armories, 55, 56
army. *See* North; Rebels; soldiers; South; Union; Yankees
army camp life, 77, 81, 82, 107–109, 110
Army of Northern Virginia, 142
arsenals, 55
artillery, 17, 84, 112, 113, 116, 126, 138. *See also* cannon; guns
assassination, 145–146, 147, 148
Atlanta, Georgia, 127–129
Auburn, New York, 33

B

ballooning, 84, 85
Baltimore, Maryland, 55
Baptist church, 35
Barton, Clara, 99, 111
baseball, 81
"Battle Hymn of the Republic," 116, 152
battles. *See* fighting; strategy;
or under name of battle
bayonets, 3, 83, 84. *See also* weapons
Beauregard, Gen. P. G. T., 18, 68, 107
Beecher, Catharine, 23, 24
Beecher, Harriet. *See* Stowe, Harriet Beecher
Beecher, Henry Ward, 23, 24, 144
Beecher, Lyman, 23, 24
Bible, 5, 32–35, 45, 54, 116, 132. *See also* religion
Bill of Rights, 149
Black Hawk War, 39, 45, 46
blacks, 12, 16, 24–26, 27–33, 43, 59, 60, 78, 103–106, 121, 139, 144, 149. *See also* abolitionism; slavery
blockade, 64, 66, 74, 94, 97, 118
bomb shelters, 118
Booth, John Wilkes, 57, 145–147
border states, 60–61, 100. *See also* Delaware, Kentucky, Maryland, Missouri
Boston, Massachusetts, 53
Bowdoin College, Maine, 25
boys, 68, 80–81, 84
Brazil, 51
bridges, 85, 126. *See also* transportation
Britain. *See* Great Britain
Brown's Indian Queen Hotel, 9, 10
Brown, Sen. Albert Gallatin, 51
Brown, John, 54–58, 147, 150
Buford, Gen. John B., 111, 112
Bull Run (Manassas), battle of, 18–22, 68, 85, 89, 99, 138, 139
bullets. *See* ammunition
Burns, Anthony, 53
Burns, John, 112
Burnside, Gen. Ambrose E., 66
Butterfield, Gen. Daniel, 84

C

cabinet, Confederate, 60, 143
cabinet, U.S., 86, 133, 143, 144, 146
Calhoun, Sen. John C., 9, 10, 13, 31, 47, 50, 53
Canada, 28, 52
cannonballs, 83, 84, 97, 112, 116, 138
cannon, 17, 18, 85, 90, 105, 113, 116, 118, 134. *See also* artillery
captains, 77, 81, 95, 116
cartridges. *See* ammunition
cavalry, 68, 69, 111, 113, 124
Chancellorsville, Virginia, battle of,
65, 71, 92–93, 109, 113
Charleston, South Carolina, 17, 18, 62, 63, 105, 138, 144
Charlestown, Virginia, 57
Chase, Salmon P., 144
Chesnut, Mary, 75, 133, 140
Chicago, Illinois, 38, 148
children, 12, 24, 27, 29, 35
Chinese, 51
Cincinnati, Ohio, 24
City Point, Virginia, 125, 126, 133, 134
Clay, Henry, 39, 52, 53
Clem, Johnny, 80
clothing, 17, 20, 21, 68, 73, 76, 111. *See also* uniforms
Columbia, South Carolina, 129
communications, 84. *See also* telegraph
Compromise of 1850, 52, 53
Confederacy, 16, 45, 47, 59, 62, 63, 64, 68, 71, 73–76, 79, 89, 103, 118, 143
Confederate army, 21, 65, 68–72, 77, 84, 105, 107, 117, 118, 125, 129, 135, 142
Confederate States of America. *See* Confederacy
Congregational church, 23
Congress, Confederate, 16, 132
Congress, United States, 10, 33, 52, 133, 144. *See also* Senate
Connecticut, 23
conscription. *See* draft
Constitution, 10, 16, 32, 49, 79, 131, 133, 139, 142, 149
constitutional amendments, 79, 133, 139, 142, 144, 149
contrabands, 104, 105
cotton, 11, 12, 13, 29, 45, 46, 47, 51, 58, 64, 73, 74, 75
crops, 11, 73, 103, 129. *See also* cotton; rice; sugar; tobacco
currency. *See* money

D

Davis, Jefferson Finis, 45–47, 49, 60, 73–75, 117, 128, 135–137, 149
Davis, Joseph, 45, 46
Davis, Varina Howell, 45, 47
deaths, 14, 83, 117, 124, 125
Declaration of Independence, 11, 43, 101, 139, 149
Delaware, 27, 60
democracy, 78–79, 148. *See also* constitutional amendments;

Emancipation Proclamation;
  Gettysburg Address
Democratic Party, 129
deserters, 92, 117, 126
disease, 82, 83, 119
Douglas, Sen. Stephen A., 41–44, 68
Douglass, Frederick, 100, 101, 103, 148
draft (conscription), 80
draft riots, 106, 120, 121
Dred Scott decision, 43, 54, 144
drummer boys, 80, 84, 103
dugouts, 118

E
earthworks, 85, 89
Eastern Shore, Maryland, 27
economics, 10, 52
elections, 42–43, 44, 128, 129, 130, 133
Emancipation Proclamation, 100–
  102, 105, 131, 133
England. *See* Great Britain
Europe, 50–52, 64, 85
Everett, Edward, 119–120, 121

F
factories, 60, 78, 79, 101, 103
Fahnestock Brothers, Gettysburg, 111
Fahnestock, Gates, 111–113, 119
farming, 11, 50, 52, 60, 62, 79, 81, 103.
  See also agriculture; crops
Farragut, Adm. David Glasgow, 76–
  77, 97
Federal Union, 9, 16. *See also* North;
  Union
Fifteenth Amendment, 139, 149
Fifty-fourth Massachusetts
  Regiment, 105
fighting, 20, 21, 66, 71, 83–85, 105–106,
  111–116, 118
firearms. *See* guns
Five Forks, battle of, 134, 135
food, 73, 81, 82, 89, 107, 118, 126,
  134, 135, 137
Ford's Theatre, 143, 144, 145
Fort Pillow Massacre, 106
Fortress Monroe, 32, 149
forts, 17, 18, 57, 62, 63, 92, 105, 106,
  107, 138, 149
Fort Sumter, South Carolina, 17, 18,
  57, 62, 63, 92, 107, 138, 144
Fort Wagner, 103, 105
Fourteenth Amendment, 139, 149

France, 50–51, 75, 85
Fredericksburg, Virginia, battle of,
  108, 109
free blacks, 26, 61
Frémont, Gen. John Charles, 66
fugitives, 29, 30, 31, 52–53
Fugitive Slave Act, 43, 52, 53, 54, 104

G
Garrison, William Lloyd, 144
Gatling, Richard, 83
generals. *See under individual names*
Georgia, 15, 45, 49, 59, 104, 127, 128
Germany, 51
Gettysburg, battle of, 111–117, 119, 135
Gettysburg, Pennsylvania, 77, 109–
  116, 118, 119–123
Gettysburg Address, 119–123
Graham, Mentor, 39
Grant, Gen. Ulysses S., 6, 64, 65, 66,
  69, 83, 105, 113, 118, 124, 125, 127,
  130, 133, 134, 137, 139, 140–145
Graybeard Regiment, 80
Great Britain, 26, 50, 52, 68, 74, 94
Greenhow, Rose O'Neal, 68
guns, 17, 18, 20, 62, 82–84, 96, 105, 106,
  112, 141. *See also* artillery; cannon;
  muskets; rifles

H
Haiti, 49, 51
Halleck, Gen. Henry W., 66
Hammond, Sen. M. B., 64
Hampton Roads, Virginia, 96, 149
Hancock, Gen. Winfield Scott, 77,
  115, 116
Hanks, Dennis, 34–37
Hanks, Nancy. *See* Lincoln, Nancy
Harper's Ferry, Virginia, 54–58
Harvard University, 87, 119
Hay, John, 120, 135
Herndon, William ("Billy"), 39
Higginson, Thomas Wentworth, 101
Hilton Head, South Carolina, 109
Hooker, Gen. Joseph, 65, 66
horses, 61, 62, 64, 65, 70, 84, 90, 112,
  115, 117, 120, 126, 141. *See also* cavalry
hunting, 29, 36
Hurricane plantation, Mississippi, 46

I
Illinois, 38–40, 41, 42, 81

Illinois General Assembly, 40
inaugural address (1861), 61
inaugural address (1865), 130–132
indentured servants, 12
Indiana, 12, 35, 34, 39, 46
Indians, 39, 141, 142. *See also* Native
  Americans
industry, 17, 52, 59, 61, 62, 63, 78, 79,
  94, 101, 103, 125
inventions, 82–84, 94, 95, 133
Iowa, 80
iron, 59, 103
ironclad ships, 94, 95, 97
Iroquois, 141, 142

J
Jackson, Andrew, 9, 10, 39
Jackson, Gen. Thomas J. "Stonewall,"
  20, 21, 57, 69–70, 71, 72, 90, 92–93, 118
James River, 89, 125, 126, 135
Jefferson, Thomas, 9, 48, 73, 101, 135
"John Brown's Body," 116, 152
Johnson, Andrew, 146, 147, 148

K
Kansas, 54
Keckley, Elizabeth, 88, 146
Kentucky, 24, 34, 35, 41, 45, 60, 61,
  88, 137
knives, 55. *See also* weapons

L
labor, 12. *See also* factories; farming;
  indentured servants; slavery
law, 40, 133
laws, 27, 32–33, 39, 40, 43, 49, 50–53
Lee, Gen. Robert E., 6, 8, 57, 65, 69,
  70–72, 91, 92, 98, 99, 108–109, 113,
  115–118, 124, 125, 132, 135, 137,
  138, 139–142, 149
letters, 129
Lincoln, Abraham, 13, 14, 16, 26, 34–
  45, 54, 58, 59–63, 65, 66, 72, 75, 82,
  86–88, 89, 90, 94, 98–102, 104, 106,
  117, 118, 119–123, 124, 128–136,
  143–145, 147, 149
Lincoln, Mary Todd, 41, 87, 88, 131,
  145, 146
Lincoln, Nancy Hanks, 34, 35
Lincoln, Robert Todd, 87, 146
Lincoln, Sarah (Abraham's sister),
  35, 36, 38

Lincoln, Sarah Bush Johnston, 36, 37
Lincoln, Thomas ("Tad"), 86, 87, 88, 134
Lincoln, Thomas (Abraham's father), 34–36
Lincoln, Willie, 86, 87, 88, 133, 146
Lincoln–Douglas debates, 42, 44
log cabins, 34, 35, 36, 45, 107
Longstreet, Gen. James ("Old Pete"), 68, 113, 115
Los Angeles, California, 77
Louisiana, 59, 64, 92, 103
L'Ouverture. *See* Toussaint L'Ouverture

M
Magruder, Gen. John, 91
Maine, 25
Malvern Hill, Virginia, 78
Manassas, Virginia, 18, 19, 138
Manassas, battle of. *See* Bull Run
Maryland, 26, 27, 31, 60, 61
Mason, George, 12
Mason, Sen. James M., 79
Massachusetts, 105, 115
McClellan, Gen. George B., 65, 66, 89–92, 98, 99, 129, 130
McLean, Wilmer, 138, 139, 141
Meade, Gen. George G., 66, 109, 112, 113, 117
medicine, 33
Merrimack (ship). *See Virginia*
Mexican War, 69, 70, 83
Michigan, 80
midshipmen. *See* navy
Minnesota, 113
Mississippi, 45, 51, 92, 118
Mississippi River, 37, 64, 97, 118, 45, 47, 51, 59
Missouri, 60, 61, 92, 105
Mobile Bay, battle of, 97
money, 63
Monitor (ship), 94, 96, 97
muskets, 83, 84, 112. *See also* guns

N
Napoleon, 20
Natchez, Mississippi, 59
Native Americans, 11, 30, 34, 141, 142. *See also* Indians
natural life, 36
navy, 66, 74, 77, 94–97, 118. *See also* blockade; ports; ships
Nevada, 132

New Orleans, battle of, 97
New Orleans, Louisiana, 37, 78, 97
New Salem, Illinois, 38–39
New York City, 105, 106
New York State, 33, 120
Nicolay, John, 120
Norfolk, Virginia, 27, 94
North, 9, 11, 12, 15–17, 21, 22, 26, 32, 47, 51, 58, 62, 64, 78, 79, 103, 108, 124, 139, 148, 149. *See also* Union army; United States; Yankees
North Carolina, 59, 127, 134
Northern army. *See* Union army

O
Ohio, 24
Ohio River, 37, 49, 60
oil, 63
Our American Cousin, 145
overseers, 24, 26

P
Palmerston, Lord, 26
Parker, Brig.-Gen. Ely, 133, 141, 142
"peculiar institution." *See* slavery
Peninsular campaign, 90–93
Pennsylvania, 63, 77, 81, 99, 109, 111–116, 119–123
Petersburg, siege of, 125, 126, 130, 134, 135
Petersburg, Virginia, 76, 125, 126, 130, 134, 135
Philadelphia, Pennsylvania, 61, 111, 144
Pickett, Gen. George, 68, 69, 111, 113–116, 135–137, 147
Pickett's Charge, 113, 114–115
pioneers, 35–36
plantations, 24, 26, 29, 46, 47, 50, 52, 134
political parties. *See* Democratic Party; Republican Party
pontoons, 125, 126
popular sovereignty, 41, 42
ports, 64, 74, 94
positive good, 10. *See also* Calhoun, Sen. John C.; slavery; states' rights
postal service, 38–39
Potomac River, 54, 85, 89, 96, 99, 117
powder store, 85
press, 57, 105, 106, 131, 136, 149
prisoners of war, 66
punishment, 27–28, 29

Q
Quakers, 28
queen of England. *See* Victoria

R
racism, 12, 16, 51, 61, 79, 103–106, 149. *See also* slavery
raiders, 84
railroads, 18, 41, 42, 52, 55, 70, 73, 84, 85, 99, 111, 125, 127, 134, 135, 138. *See also* transportation
rebellion, 48, 49, 148. *See also* Brown, John; Toussaint L'Ouverture; Turner, Nat
Rebels, 16, 18, 21, 62, 63, 68, 77, 79, 99, 103, 111–117, 126, 128, 134. *See also* Confederacy; Confederate army; South
Reconstruction, 131, 143, 148
regiments, 81, 89
religion, 32, 48, 50, 54–58, 70, 71. *See also* Bible; Congregational Church; Quakers; religious faith
religious faith, 23, 27, 69
reporters. *See* press
Republican Party, 42, 44
revolts. *See* rebellion
Revolutionary War, 11, 12, 16, 20, 83, 103
Rhodes, Pvt. Elisha, 65, 92, 99, 146
rice, 11, 73. *See also* crops
Richmond, Virginia, 18, 21, 59–60, 68, 89, 91, 92, 93, 124, 125, 126, 130, 134–137
rifles, 82–83. *See also* guns
riots. *See* draft riots
River Queen (ship), 134, 143
rivers. *See* transportation
roads, 127. *See also* transportation
Rock, John Swett, 144
Ruffin, Edmund, 57
runaways. *See* fugitives

S
sailors. *See* navy
Santo Domingo. *See* Haiti
Sauk and Fox Indians, 39
Savannah, Georgia, 20
schooling, 24, 29, 35, 39, 46
Scott, Gen. Winfield, 64–65
secession, 14, 15, 16, 44, 59, 60, 71. *See also* states' rights

Semmes, Capt. Raphael, 95
Senate, U.S. 47. *See also* Congress, U.S.
Seneca Indians, 141
Seward, William Henry, 32
Sharps, Christian, 82
Sharpsburg, Maryland, 98, 109. *See also* Antietam, battle of
sharpshooters, 83, 116, 106
Sharps rifle, 82. *See also* rifles
Shaw, Col. Robert Gould, 105
Shenandoah River, 54
Shenandoah Valley, 70, 90, 91, 129
Sheridan, Gen. Philip, 129, 134
Sherman, Gen. William Tecumseh, 66, 126–129, 134, 143, 149
Sherman's March, 127–129
shinplasters, 63
ships, 64, 66, 68, 73, 94–97
shoes, 111
slave catchers, 27, 30, 31
slavery, 9–13, 15, 22, 24–33, 35, 40, 41, 42, 43, 46, 48–53, 54, 56, 58, 61, 62, 66, 73, 74, 75, 78, 79, 100–106, 120, 129, 130, 131, 132, 144, 149. *See also* abolitionism; blacks; South
slave states, 12, 16, 24, 43, 60, 61
slave trade, 49–50
Small, Robert, 106
snipers, 116. *See also* sharpshooters
soldiers, 17, 20, 21, 62, 68–70, 76–77, 80–85, 103–109, 149. *See also* camp life; Confederate army; Union army
songs, 80, 84, 101, 116, 150–152
South, 9, 11, 12, 15–17, 21, 22, 26, 43, 47–49, 51, 52, 58, 62, 64, 78, 79, 103, 108, 126, 127, 129, 130–131, 139, 143, 147–149. *See also* Confederacy; Confederate army; Rebels
South Carolina, 9–10, 32, 57, 58, 59 63, 18, 78, 32, 109, 127, 129, 138, 144
Southern army. *See* Confederate army
Speed, Joshua, 38
spies, 68, 84
sports, 81
Springfield, Illinois, 40, 44, 54, 87, 148
Stanton, Edwin M., 65, 96
states' rights, 10, 16, 62, 71, 74, 75. *See also* secession
Stephens, Alexander, Confederate vice president, 16
St. Louis, Missouri, 61
Stowe, Harriet Beecher, 23, 25, 26, 27, 61, 74, 144. *See also Uncle Tom's Cabin*
strategy, 20, 64–65, 83, 84, 90–91, 108, 115, 118, 124, 125, 127, 137. *See also* Anaconda Plan; Peninsular campaign; Sherman's March; total war
Stuart, Gen. J. E. B., 68, 69, 76, 113
submarines, 84
sugar, 11, 73. *See also* crops
Sumner, Sen. Charles, 144, 146
Supreme Court, U.S., 43, 54, 144
surrender, 138–142
swords, 3, 55, 139, 140. *See also* weapons

T

"Taps," 84
Taney, Roger B, 144
Taylor, Gen. Zachary, 46, 47
technology. *See* inventions
telegraph, 84, 133
Tennessee, 59, 92, 127
tents, 107. *See also* camp life
Texas, 59, 64, 103
Thirteenth Amendment, 139, 149
Thomas, Gen. George H., 77
Thoreau, Henry David, 33
tobacco, 11, 12, 73, 74. *See also* crops
Tocqueville, Alexis de, 49
total war, 85, 108, 128–129
Toussaint L'Ouverture, 49
trading, 64, 73, 74, 75
training, 81–82, 89
transportation, 18, 41, 42, 52, 55, 70, 73, 84, 85, 99, 111, 117, 125, 127, 135, 138. *See also* railroads; roads; ships
Traveller, 70, 72
trenches, 84, 127
Tubman, Harriet, 27–33
Turner, Nat, 48, 54

U

Uncle Tom's Cabin, 24, 25, 26, 61, 74, 75. *See also* Stowe, Harriet Beecher
Underground Railroad, 28–32. *See also* Tubman, Harriet
uniforms, 17, 20, 21, 68, 76, 81
Union army, 11, 57, 64–67, 72, 77, 89, 90, 103–107, 109, 126, 127
Union, 9, 10, 13, 15, 16, 32, 47, 61, 63, 76, 79, 79, 100, 112, 118, 143, 148. *See also* Federal Union; North
United States, 16, 120, 131, 136, 148. *See also* North; Union
U.S. Army of the Potomac, 65, 126
U.S. Army of the West, 127

V

Vermont, 50
Vicksburg, Mississippi, 97, 118, 125
Victoria, queen of England, 33, 75
Virginia, 27, 48, 54, 59, 60, 70, 72, 76–79, 89, 90, 94, 99, 108, 117, 129, 138, 145, 147
Virginia (*Merrimack*, ship), 95, 96, 97
Virginia Bill of Rights, 12
Virginia General Assembly, 48–49
Virginia Military Institute, 69, 93
Virginia Peninsula, 89
volunteers, 22, 59, 80, 84
voting, 139, 149

W

war casualties. *See* deaths; wounded
warfare. *See* fighting; strategy
warships. *See* navy; ships
Washington, D.C., 18, 21, 55, 68, 88, 90, 91, 117, 120, 124, 133, 135, 138, 144, 148
Watkins, Pvt. Sam, 20, 71, 80, 149
weapons, 3, 17, 18, 54, 55, 59, 62, 64, 82–84, 90, 94, 96, 105, 106, 112, 139, 140, 141
western territories, 12, 13, 41, 47, 62, 129
West Point, U.S. Military Academy, 17, 46, 65, 66, 68–70, 78, 83, 127
West Virginia, 60
White House of the Confederacy, 60, 135, 136
White House, Washington, D.C., 86, 87, 88, 120, 145, 146
Whitman, Walt, 72, 136, 146
Wilderness, battle of, 92, 124, 125, 126
Williamsburg, Virginia, battle of, 89
wounded, 14, 117, 119, 124, 149

Y

Yankees, 16, 17, 47, 52, 60, 62, 68, 77, 108, 109, 112, 115, 117, 119, 125, 126, 135. *See also* North, Union army
York River, 89
Yorktown, Virginia, 89

Z

Zouave soldiers, 20

Picture Credits

Cover: Kurz & Allison, *Battle of Antietam*, collection of Walter Lord; **5**: William L. Sheppard, *Under Fire, Petersburg*, American Heritage Publishing Company; **6**: Library of Congress; **6–7 (top)**: Vermont Historical Society, Montpelier; **6–7 (bottom)**: U.S. Army Military History Institute, Carlisle Barracks, Pennsylvania; **8**: Library of Congress; **9 (left)**: National Portrait Gallery, Smithsonian Institution; **9 (right)**: New York Public Library; **10**: Culver Pictures; **11 (left)**: Peabody Museum, Harvard; **11 (right)**: Library of Congress; **12 (left)**: Chicago Historical Society; **12 (right)**: Library of Congress; **13**: National Archives; **14 (top)**: *Charlestown Courier*, September 10, 1861; **14 (bottom), 15**: Library of Congress; **16**: Chicago Historical Society; **17 (top)**: National Archives; **18 (top)**: Library of Congress; **18 (bottom)**: National Archives; **20 (left)**: Library of Congress; **21 (top)**: *Battles and Leaders of the Civil War*, Library of Congress; **22**: Library of Congress; **24**: Stowe-Day Foundation, Hartford, Connecticut; **25 (top)**: Chicago Historical Society; **25 (bottom)**: Library of Congress; **25 (left, right)**: Warshaw Collection of Business Americana; **26**: Library of Congress; **27, 28 (top)**: Library of Congress; **28 (bottom)**: Metropolitan Museum of Art; **29**: National Archives, records of the War Department, general and special staffs; **30**: Library of Congress; **31 (top)**: American History Picture Library; **31 (bottom)**: Library of Congress; **32 (left)**: National Portrait Gallery, Smithsonian Institution; **32-33**: Huntington Library; **34 (top)**: Library of Congress; **34 (bottom)**: Lloyd Ostendorf, Dayton, Ohio; **35**: Chicago Historical Society; **36**: Illinois State Historical Society, Springfield; **37**: Library of Congress; **38 (top)**: Library of Congress; **38 (bottom)**: Isaac Hillard and the Speed Museum, Farmington; **39 (top)**: Library of Congress; **39 (inset)**: Lloyd Ostendorf, Dayton, Ohio; **40**: Library of Congress; **41 (left)**: Chicago Historical Society; **41 (right), 42 (left, right)**: Library of Congress; **42 (middle)**: National Portrait Gallery, Smithsonian Institution; **44 (top)**: Chicago Historical Society; **44 (bottom)**: *Frank Leslie's Illustrated Newspaper*; **45**: National Portrait Gallery, Smithsonian Institution; **46 (top left)**: Museum of the Confederacy, Richmond, Virginia; **46 (bottom left)**: Library of Congress; **46 (top right)**: Mississippi Department of Archives and History, Jackson; **46 (bottom right)**: U.S. Army Military History Institute, Carlisle Barracks, Pennsylvania; **47 (top)**: Library of Congress; **48 (top)**: Peabody Museum, Harvard University; **48 (bottom)**: Library of Congress; **49**: Mansell Collection; **50 (top)**: Huntington Library; **50 (bottom)**: Library of Congress; **51**: New-York Historical Society; **52**: "The Land of Liberty," *Punch*, 1847; **53**: Library of Congress; **54 (top)**: Library of Boston Athenaeum; **55 (top)**: Chicago Historical Society; **55 (bottom)**: Harper's Ferry National Historical Park; **56, 57(left)**: Library of Congress; **57 (right), 58 (left)**: Chicago Historical Society; **59 (top)**: Mellon Collection; **59 (bottom)**: Connecticut Historical Society; **60 (left)**: *Harper's Weekly*; **60 (right)**: U.S. Army Military History Institute, Carlisle Barracks, Pennsylvania; **60 (bottom)**: Virginia State Library; **61**: Library of Congress; **63 (bottom)**: Drake Well Museum, Titusville, Pennsylvania; **64, 65 (left)**: Library of Congress; **66 (top)**: Western Reserve Historical Society; **66 (bottom)**: Library of Congress; **68 (top)**: Larry B. Williford, Portsmouth, Virginia; **68 (bottom left)**: David L. Hack; **68 (bottom right), 69, 70 (right)**: Library of Congress; **70 (left)**: Washington/Custis/Lee Collection, Washington and Lee University, Lexington, Virginia; **71 (top)**: National Archives; **71 (bottom)**: Casemate Museum.; **72**: Library of Congress; **73 (top)**: R. W. Norton Art Gallery, Shreveport, Louisiana; **73 (bottom)**: National Archives; **74 (top)**: Library of Congress; **74 (bottom)**: Lightfoot Collection; **75 (top)**: National Portrait Gallery, Smithsonian Institution; **75 (bottom)**: Library of Congress; **76 (top)**: John A. Hess; **76 (bottom)**: Library of Congress; **77 (left)**: Cook Collection, Valentine Museum, Richmond, Virginia; **77 (right)**: U.S. Army Military History Institute, Carlisle Barracks, Pennsylvania; **78 (top right)**: U.S. Army Military History Institute, Carlisle Barracks, Pennsylvania; **78 (bottom)**: Library of Congress; **79 (top)**: Sterling Chandler and Richard S. Young; **79 (bottom)**: *Punch*, May 18, 1861; **80 (top)**: Library of Congress; **81 (top)**: John A. Hess; **81 (bottom)**: Library of Congress; **82 (left)**: Lightfoot Collection; **83 (top, bottom)**: Chicago Historical Society; **83 (middle)**: *Heroes of the Civil War* and *Generals and Battles of the Civil War*; **84, 85 (top left, bottom)**: Library of Congress; **85 (top right)**: U.S. Army Military History Institute, Carlisle Barracks, Pennsylvania; **86 (top)**: Lincoln Museum; **87 (top)**: David L. Hack; **87 (right)**: Illinois State Historical Library, Springfield; **87 (bottom)**: Library of Congress; **88 (top)**: Louise and Barry Taper Collection; **88 (bottom)**: *Frank Leslie's Illustrated Newspaper*; **89**: Amon Carter Museum; **90**: Library of Congress; **91**: *Heroes of the Civil War* and *Generals and Battles of the Civil War*; **92-93**: Library of Congress; **94**: U.S. Army Military History Institute, Carlisle Barracks; **95, 96 (top)**: Library of Congress; **96 (bottom)**: *Heroes of the Civil War* and *Generals and Battles of the Civil War*; **97 (right)**: Chicago Historical Society; **97 (left)**: National Archives; **98, 99**: Library of Congress; **100 (top)**: Huntington Library; **100 (bottom)**: National Portrait Gallery, Smithsonian Institution; **101, 102**: Chicago Historical Society; **103 (top)**: John A. Hess; **103 (bottom), 104 (bottom left)**: U.S. Army Military History Institute, Carlisle Barracks, Pennsylvania; **104 (top)**: Library of Congress; **104 (bottom right)**: *Heroes of the Civil War* and *Generals and Battles of the Civil War*; **105, 106 (top left)**: Library of Congress; **106 (top right)**: Illinois State Historical Society, Springfield; **106 (bottom right)**: U.S. Army Military History Institute, Carlisle Barracks, Pennsylvania; **106 (bottom)**: Naval Historical Center, Washington, D.C.; **106 (inset)**: U.S. Army Military History Institute, Carlisle Barracks, Pennsylvania; **107 (top)**: U.S. Army Military History Institute, Carlisle Barracks, Pennsylvania; **107 (bottom)**: Robert McDonald; **108–109**: International Center of Photography, George Eastman House, Rochester, New York; **110 (top)**: Lightfoot Collection; **110 (bottom)**: Library of Congress; **111 (top)**: Cook Collection, Valentine Museum, Richmond, Virginia; **111 (bottom)**: Gettysburg National Military Park; **112 (top)**: Library of Congress; **112 (bottom)**: U.S. Army Military History Institute, Carlisle Barracks, Pennsylvania; **113**: Lee-Fendall House, Alexandria, Virginia; **115**: Library of Congress; **116 (top)**: Chrysler Museum; **116 (bottom)**: U.S. Army Military History Institute, Carlisle Barracks, Pennsylvania; **117 (top)**: Library of Congress; **118, 119 (left)**: Library of Congress; **120**: Lloyd Ostendorf, Dayton, Ohio; **121 (left)**: Lightfoot Collection; **121 (right)**: Lincoln Museum; **122–123 (top)**: U.S. Army Military History Institute, Carlisle Barracks, Pennsylvania; **122 (bottom), 123 (right)**: Library of Congress; **124 (top)**: T. K. Treadwell Collection; **124 (bottom)**: Library of Congress; **127 (bottom)**: records of the Office of the Chief Signal Officer; **128 (bottom)**: *Frank Leslie's Illustrated Newspaper*; **129 (top)**: Hallmark Photographic Collection; **129 (bottom)**: Lincoln Museum; **131 (right)**: *Harper's Weekly*; **131 (left)**: Library of Congress; **132 (top)**: Huntington Library; **132 (bottom)**: Lloyd Ostendorf, Dayton, Ohio; **133 (top)**: Library of Congress; **133 (bottom)**: U.S. Army Military History Institute, Carlisle Barracks, Pennsylvania; **134 (top)**: Library of Congress; **134 (bottom)**: *Heroes of the Civil War* and *Generals and Battles of the Civil War*; **135**: Chicago Historical Society; **136 (top)**: Library of Congress; **137, 138, 139 (top)**: Library of Congress; **139 (bottom)**: Chicago Historical Society; **141 (right)**: Appomattox Court House National Historical Park; **142**: Library of Congress; **143 (top)**: Chicago Historical Society; **143 (bottom), 145 (top)**: Library of Congress; **145 (middle left)**: West Point Museum, U. S. Military Academy; **145 (middle right)**: Huntington Library; **145 (bottom)**: Chicago Historical Society; **146 (top)**: Library of Congress; **147 (top)**: Huntington Library; **148 (top, bottom)**: Library of Congress; **148 (middle)**: Chicago Historical Society; **149 (right)**: William Gladstone Collection; **159**: *Harper's Weekly*; **160**: Chicago Historical Society

A Note From the Author

A family remembers one Civil War soldier who left them and his horse behind—and did not come home again.

Let's be mice peeking through a tiny hole in the walls of a comfortable old house: a house with tasty crumbs on the floor, good music, and happy people.

Look at that family sitting around the dinner table, telling stories, joking, and laughing. Father is a carpenter, and a poet; when people read his poetry it makes them smile, and think, and sometimes wipe away a tear, too. Mother plays the violin, sews the children's clothes, and is the heart of this busy 19th-century household. All the children love music.

Sally plays piano, Tom the flute, Harry a horn, and Mary the cello. Father sings. Tonight they will have a family concert. Afterwards they will have cookies and cake, and after that, when everyone is asleep, we mice will have a party.

That was all pretend. It wasn't real at all. Never. The father you saw was killed at the battle of Bull Run. He was 18. He never married. He never wrote the poetry he would have written if he had lived. The mother you saw never married either. Most of the young men in her town didn't come back from the war, so there was no one to marry. Besides, she loved the young man who died at Bull Run. Of course, the four children were never born.

And that may be the worst part of war: the killing of the young. The killing of a nation's potential. It is often the brightest and bravest who die. Lost forever are the poems they might have written, the symphonies they might have composed, and the children they might have borne.

> *I have never had a feeling, politically, that did not spring from…the Declaration of Independence. I have often inquired of myself what great principle or idea it was that kept…[us] so long together. It was not the mere matter of separation of the colonies from the motherland, but that sentiment in the Declaration of Independence which gave liberty not alone to the people of this country, but hope to all the world for all future time. It was that which gave promise that in due time the weights would be lifted from the shoulders of all men, and that* all should have an equal chance. *This is the sentiment embodied in the Declaration of Independence.* —Abraham Lincoln

When you are writing history, it is hard to end a book. I always worry that I haven't explained things well enough. And I always know that there is more to tell. Ending this book is especially difficult.

Poet Walt Whitman (who worked as a nurse during the war) said, "Future years will never know the seething hell and the black infernal background, the countless minor scenes and interior of the secession war; and it is best that they should not. The real war," said Whitman, "will never get in the books."

Maybe Whitman is right. Maybe words can never explain war—although many of us use them to try it. No war—well, certainly no American war—has been more written about than the war between the states. Some 50,000 books have been published about those years when we fought against our brothers and sisters. Why are we so interested in the Civil War? Could it be that in that war lies the purpose of our nation? What we truly believe; what a generation was willing to die to preserve.

We weren't fighting for land. After the first awful battle at Bull Run, South Carolina soldier J. W. Reid wrote, "We are now occupying the same ground that we did before the battle."

It was the same after the war was over. We were back where we had begun five years before. We were occupying the same ground. But we weren't the same. That idea in the Declaration of Independence had won. It was now clear: in America all were meant to "have an equal chance"; that was the meaning behind this war. It was the meaning of America. But it would take more tears, more energy, and new kinds of battles to begin to make it happen.